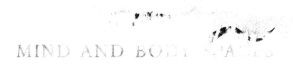

MIND AND BODY SPACES

Geographers are increasingly engaged with both the theoretical debates surrounding ill or impaired bodies and the lived realities of ill/impaired experience. Just as geographies of race, gender, class and sexuality have drawn attention to how complex power relations in society are spatialised, so geographies of illness and impairment offer a deeper understanding of the world.

Mind and Body Spaces highlights new international research – from Britain, USA, Canada and Australia – on bodily impairment, mental health and disabling social worlds. A range of different spatial 'settings' in which different minds and bodies are always located are examined, including the nation, urban and rural spaces, work spaces, the 'caring' institution, the street and the home.

The contributors discuss varied issues concerning physical impairment and mental health, ranging from historical conceptions of the body and behaviour to contemporary political activism. This range of concerns also includes matters of identity and employment, accessible housing, parenthood and child carers, psychiatric medication use, masculinity, sexuality, autobiography, social exclusion and inclusion.

In a deliberate attempt to extend conventional geographical research concerning disability, this collection clearly illustrates the complex interconnections between mind/body states and wider socio-cultural, economic, political and medical environments. Bringing together entities traditionally kept apart – mind and body, illness and impairment – this book seeks to invigorate debate about the diverse geographies of ableism.

Ruth Butler is Lecturer in Applied Social Research at The University of Hull and **Hester Parr** is Lecturer in Geography at The University of Dundee.

D0232189

CRITICAL GEOGRAPHIES
Edited by Tracey Skelton
Lecturer in International Studies, Nottingham Trent University
and
Gill Valentine
Senior Lecturer in Geography, The University of Sheffield.

This series offers cutting-edge research organised into three themes: concepts, scale and transformations. It is aimed at upper-level undergraduates and research students, and will facilitate inter-disciplinary engagement between geography and other social sciences. It provides a forum for the innovative and vibrant debates which span the broad spectrum of this discipline.

1. MIND AND BODY SPACES
Geographies of illness, impairment and disability
Edited by Ruth Butler and Hester Parr

2. EMBODIED GEOGRAPHIES
Spaces, bodies and rites of passage
Edited by Elizabeth Kenworthy Teather

3. LEISURE/TOURISM GEOGRAPHIES
Practices and geographical knowledge
Edited by David Crouch

4. CLUBBING
Dancing, ecstasy, vitality
Ben Malbon

5. ENTANGLEMENTS OF POWER
Geographies of domination/resistance
Edited by Joanne Sharp, Paul Routledge, Chris Philo and Ronan Paddison

6. DE-CENTRING SEXUALITIES
Politics and representations beyond the metropolis
Edited by Richard Phillips, Diane Watt and David Shuttleton

MIND AND BODY SPACES

Geographies of illness, impairment and disability

Edited by
Ruth Butler and Hester Parr

London and New York

First published 1999
by Routledge
11 New Fetter Lane, London EC4P 4EE

Simultaneously published in the USA and Canada
by Routledge
29 West 35th Street, New York, NY 10001

Routledge is an imprint of the Taylor & Francis Group

© 1999 Ruth Butler and Hester Parr for editorial and selection;
individual chapters, the contributors

Typeset in Perpetua by
The Florence Group, Stoodleigh, Devon
Printed and bound in Great Britain by
T.J. International Ltd, Padstow, Cornwall

British Library Cataloguing in Publication Data
A catalogue record for this book is available from the British Library

Library of Congress Cataloging in Publication Data
A catalogue record is available from the Library of Congress

ISBN 0–415–17902–5 (hbk)
ISBN 0–415–17903–3 (pbk)

CONTENTS

CONTENTS

ILLUSTRATIONS

Plates

Figures

CONTRIBUTORS

Ruth Butler is a lecturer in Applied Social Research in the School of Comparative and Applied Social Sciences, University of Hull, where she teaches a course on disability policy, identity and society. She has written on the oppression and resistance of people with disabled bodies in general, and visually impaired people and disabled youths in particular. She is currently working with Tracey Skelton and Gill Valentine on an ESRC-funded project on the experiences of deaf, gay and lesbian youths.

Vera Chouinard is Professor of Geography in the School of Geography and Geology, McMaster University, Hamilton, Canada. Her research is concerned with state intervention in cities, political and legal regulation of urban struggles and alternative services, and the role of differences such as class, gender and disabilities in people's capacities to struggle for social change. Her current research projects include socio-spatial factors influencing disabled women's experiences of activism in Canada and the impacts of state restructuring on disabled people's lives and capacities to contest oppression.

Michael L. Dorn is a doctoral candidate in Geography at the University of Kentucky. His work addresses socio-medical geographies of physical difference (especially disability, race and gender), the historical geography of medical environmental science, and social theories of scientific rationality and the body.

Isabel Dyck is an associate professor in the School of Rehabilitation Sciences, University of British Columbia. Her research interests in the area of geographies of disability focus on feminist analyses of the experiences of women with chronic illness. Other current research concerns integration issues for immigrants to Canada, including family reconstitution and immigrants' use of traditional and biomedical health care systems. Recent publications include 'Women with disabilities and everyday geographies: home space and the contested body' in R. A. Kearns and W. Gesler (eds) *Putting Health into Place: Landscape, Identity and*

Wellbeing, Syracuse: Syracuse Press and 'Methodology on the line: constructing meanings about "cultural difference" in health and health care research' in S. Grace, V. Strong-Boag, J. Anderson and A. Eisenberg (eds) *Painting the Maple: 'Race', Gender and the Construction of Canada*, UBC Press.

Flora Gathorne-Hardy is a PhD student at the Department of Geography, University of Cambridge. Her thesis examines issues of social justice and the politics of decision-making about the design of affordable housing, with comparative research being carried out in the UK and the US. This work has been funded by the Economic and Social Research Council. Her research interests revolve around the relationships between people's experiences of social injustice and the geography of the built environment.

Brendan Gleeson is presently a research fellow in the Urban Research Program at the Australian National University. His research interests centre on urban social policy, environmental policy and spatial regulation. His most recent book, *Geographies of Disability*, was released in early 1999.

Edward Hall is a research fellow in the Faculty of Social Sciences at The Open University. His research centres on issues of disability, embodiment and employment. He has recently completed a PhD thesis on disability and flexible employment and is currently researching the impact of the UK Disability Discrimination Act (1995). He has presented at several UK and international conferences.

Rob Imrie is Reader in Human Geography at Royal Holloway University of London. At present he is directing an ESRC-funded project on property markets and disabled people's access needs in Sweden and the UK.

Christine Milligan is currently a teaching fellow in the Department of Geography at the University of Strathclyde. Her principal research interests are focused around the social and spatial manifestations of health and social care restructuring – in particular for community-based individuals with mental ill-health and the frail elderly. Her research has recently been published in *Health and Place* (1996) and *Social Science and Medicine* (1998). She is currently completing her doctoral thesis, which focuses on the role of informal care providers to community-based groups in the Scottish context.

Pamela Moss (Associate Professor) teaches as a feminist geographer at the University of Victoria, Victoria, BC, Canada. Her interests in commonplace activities and the mundane has led her to explore the everyday lives of low-income women, older women living with arthritis, and women diagnosed with chronic illness. She is also active in feminist community politics.

Deborah Carter Park received her PhD in Geography from York University in 1995. She is currently studying at the Ontario Institute for Studies in Education of the University of Toronto. Her research interests are in the field of intellectual disability. She has published papers in the *Journal of Historical Geography*, *Annals of the Association of American Geographers* and, more recently, in *Progress in Human Geography* and *Disability and Society*.

Hester Parr is a lecturer in Human Geography at Dundee University. Her research work is concerned with the deinstitutionalisation of people with mental health problems in the UK. In particular this work has looked at issues of identity, corporeality, geographical imagination, the experience of everyday life and the collective and individual ways in which people with mental health problems resist psychiatric inscription. Her work has been published in *Area, Environment and Planning D: Society and Space* and *Health and Place*. She is currently working on questions of mental health in rural Highland communities.

John Radford is Associate Professor of Geography at York University, Ontario, Canada. He has written on society and space in the nineteenth-century city, especially in the southern United States, and on the asylum, including articles in *Social Science and Medicine, Journal of Historical Sociology* and, more recently, in *Progress in Human Geography* and *Disability and Society*.

Fiona Smith is a lecturer in the Geography and Earth Sciences Department of Brunel University. She is also the director of the 'Out of School Childcare Research Unit' in the department. The unit is currently undertaking numerous research projects including the 'Child centered after-school and holiday childcare' project, which is funded by the ESRC. Fiona has published in *Children's Environments, Built Environments* and is author (along with John Barker) of *The Expansion of Out of School Care* (Kids Club Network, London).

Jane Stables is a teacher of English in a secondary school in Hampshire, England.

Gill Valentine is a senior lecturer in Geography at the University of Sheffield where she teaches social geography, approaches to human geography and qualitative methods. Her research interests include: geographies of children, youth and parenting; food and foodscapes and geographies of sexualities. She is co-author, with David Bell, of *Consuming Geographies* (Routledge, 1997), co-editor, with David Bell, of *Mapping Desire: Geographies of Sexualities* (Routledge, 1995) and co-editor, with Tracey Skelton, of *Cool Places: Geographies of Youth Cultures* (Routledge, 1998).

PREFACE AND ACKNOWLEDGEMENTS

This edited collection emerged out of an IBG/RGS session on 'disability and space' which was sponsored by the Women and Geography Study Group and the Social and Cultural Geography Research Group of the IBG/RGS. We would like to thank all the participants for their papers in the session and we hope that some of the spirit of that day is translated onto these pages. The study groups were very supportive of our attempts to discuss these issues from non-medical perspectives, and we thank both groups for the financial assistance for two overseas speakers to attend the conference. It should be noted that this collection has deliberately sought the inclusion of work from established academics, young researchers and post-graduate students. We also want to acknowledge the enthusiastic and helpful assistance of Chris Philo, and the supportive and useful comments of our referees, Sophie Bowlby and Robyn Longhurst. Special mention goes to Andrew Black of Dundee Geography Department who painstakingly translated the disks of the authors and made Hester's life a lot easier. Sarah Carty and Sarah Lloyd of Routledge have been patient and helpful with the final editing arrangements of the book.

Other acknowledgements

We would like to thank members of the Young Disabled People's Project, Chester for permission to reproduce the figures in Chapter 11. 'Young Disabled People Do It Too' is a series of cartoons which explore issues around, youth, sex and disability. They were produced in 1998 by the Young Disabled People's Project at DIAL House Chester, in conjunction with Cheshire Health Promotion and cartoonist Angela Martin. The images are available to buy in either postcard or poster format (cheques payable to 'YDP CHESTER') from: Young Disabled People's Project, DIAL House, Hamilton Place, Chester CH1 2BH, Tel: 01244 315025 (e-mail: ydpchester@compuserve.com). Set of five postcards £1.00 (inc p + p)*. Big 'A2' poster £2.00 (inc p + p)*. NB* overseas orders please add £2.00 per order.

The case study 'Nine years old and looking after mum' was first published in accompaniment to the article 'Child carers need help themselves' by Glenda Cooper in *The Independent* 3 August 1995 and is reproduced in Chapter 14 with their kind agreement.

1

NEW GEOGRAPHIES OF ILLNESS, IMPAIRMENT AND DISABILITY

Hester Parr and Ruth Butler

Introduction

I've found that if I don't have a good living situation, I'll end up back in hospital. I'm serious when I tell people I get suicidal if I don't have a good place to live.

(quoted in Kearns 1987: 165–6,
cited in Taylor *et al*. 1989: 153)

I get up at 8 o'clock in the morning and I usually have coffee and watch Canada AM. And then I do my exercises . . . and I putter around the house or . . . I seem to have a heck of a lot of medical appointments. I don't think I've had a week without a medical appointment of some description.

(quoted in Dyck 1995: 316)

Since I had my cataracts operated on, that's given me, the cosmetic effect of it has given me more self confidence. Because I look more normal, or so called 'normal'.

(quoted in Butler and Bowlby 1997: 423)

The three extracts above are from interviews quoted in three recent human geographical studies: the first being orientated around people with mental health problems, the second around women with multiple sclerosis and the third around people with visual impairments. These three quotations, and the studies from which they are drawn, are linked in various and quite problematic ways. At first glance these studies are dealing with very different aspects of the mind and body (in biomedical terminology: mental illness, multiple sclerosis and

1

blindness), and very different sorts of social and spatial concerns. Attention to different embodied experiences such as these might have traditionally been regarded as either irrelevant to human geographical enquiry or as quite separate concerns, relegated to the separate sub-disciplinary fields of the 'geography of mental health', 'medical geography' and a 'geography of disability'. A key purpose of this collection is precisely to explore the interconnections between the people quoted above (amongst others), and to provide the starting point for considering together the various (and so far quite discrete) strands of human geographical enquiry which have encountered the often limited, painful, confused, restricted and segregated mind and body. The claim made in this introduction is that there are many points of connection between very different people rooted in their shared experiences of biomedical inscription, pain, social isolation and political and economic marginalisation. Moreover, we also claim that such connections exist with respect to positive experiences of community, political action and embodied resistance to stereotypical labels of illness, impairment and disability.

In order to explore these connections more fully in this chapter, we first set out the two basic 'models' (social and medical) which inform disability studies. Then, taking this characterisation in the next section, we set out to identify how the discipline of human geography has (at worst) remained ignorant of or has (at best) separated out these social and medical aspects of human existence: it has in effect fragmented the ill, impaired and different body and mind. By acknowledging the problems and possibilities of naming and aligning such expe-riences as mental health problems, chronic illness and physical impairment, we seek to destabilise what we see both as a limiting mind–body dualism in human geography and a reluctance fully to reveal or to critique the myth of the perfect mind/body. We go forward to argue that discrete entities of the discipline have much to gain by drawing on each other, and on recent advances in social theory which seek to recognise embodied difference, while also noting that we wish to avoid what Dorn has called the 'poststructurist romance' of nomad thought and the hybrid body (Dorn 1998: 184). This refers to how we draw back from just celebrating bodily/psychic diversity in this volume: we seek to recognise the genuine problems and challenges which can attach to having bodies/psyches that are different from mainstream 'norms', and as Dorn argues, we do not wish 'to exemplify a flight from the messiness of disability into myth and metaphor' (Dorn 1998: 184). Following this we consider how the 'new medical geography' can help us to conceptualise connections between the different people represented in the above quotations, and, finally, the chapter draws to a close by commenting briefly on the fourteen chapters in the collection and arguing for a reinvigorated effort to explore the diverse geographies of ill, impaired and different minds and bodies.

Existing approaches to disability and chronic illness

The two models in disability and chronic illness studies

On a very basic level, the debates about people with various impairments can be seen to have revolved around two conceptual departure points: the social and the medical models of disability. Taking the latter category first, it is possible to claim that traditionally, especially in Western societies, impairments have been seen as 'individual medical tragedies' (Shakespeare 1993) in which the body is conceptualised as simply failing to meet normal standards of form, ability and mobility (thus often incurring varied medical diagnoses and treatments in medical–institutional geographies). The assumption here is that the individual body which is at fault can be treated by largely medical interventions and technologies. There are many different theories and texts which deal critically with the ideas and practices of medicalisation (Finklestein 1980; Foucault 1967; Oliver 1990; Turner 1987; Zola 1972), and these are too numerous to develop here. Brisenden (1986 cited in Oliver 1990: 49) is useful, though, in providing a view of the difficulty of seeing body and mind differences as purely medical problems:

> The medical model of disability is one rooted in an undue emphasis on clinical diagnosis, the very nature of which is destined to lead to a partial and inhibiting view of the disabled individual. In order to understand disability as an experience, as a lived thing, we need more than the medical 'facts' . . . The problem comes when they determine not only the form of treatment (if treatment is appropriate), but also the form of life for the person who happens to be disabled.
>
> (Brisenden 1986: 173)

The medical model of disability then has been seen as a mechanism by which people with mind and body differences have been categorised and responded to by a Western society which increasingly pathologises certain peoples, and is particularly associated with the rise of medical science. Research undertaken from a medical model perspective has had some value, often in relation to the development of specific bodily and sensory 'technologies' which can improve aspects of daily life for some people with impairments. Research from this perspective has also, on occasion, involved some limited recognition of the role society plays in constructing the problems of people with impairments as 'disabled'. However, such a recognition has rarely extended to a full questioning of the social categorisation of 'disability' or the complexity of disabled people's lives.

In recent years disability theorists have put forward a new social model of disability which attempts to address these issues, and which has been closely allied to political fights for anti-discrimination legislation and civil rights. At the

3

centre of the new social model have been useful definitions of impairment and disability.

> Impairment – Lacking all or part of a limb, or having a defective limb, organism or mechanism of the body.

> Disability – The disadvantage or restriction of activity caused by a contemporary social organisation which takes no or little account of people who have physical impairments and thus exclude them from the mainstream of social activities.

> (UPIAS 1976)

Although mainly concerning the physical body (and hence only partially useful for our purposes), these definitions seek to emphasise that debates about and actions surrounding 'disability' should shift away from concentrating on physiology in order to view more critically the society in which individuals and their impairments are contextualised. Disability, then, has been said to be 'socially constructed' and society's role in constraining disabled people's lives recognised (Barnes 1991; Barton 1996; Oliver 1984, 1990; Swain *et al.* 1994). Amongst other things, the social model has emphasised the importance of broad economic structures of society in creating disability (Barnes 1991; Barton 1996; Oliver 1990; Shakespeare 1993; Swain *et al.* 1994) by devaluing bodies which do not straightforwardly conform to the time–space work regimes of capitalist society. Such an approach recognises that it is society's organisation which 'disables' people with physical and/or mental limitations so that they are marginalised socially, economically and politically. In recognising these factors the social model arguably rejects physiological and psychological difference as a reductionist explanation for inequality. The model has thus been invaluable as a basis for the critical mobilisation of disability movements as common features of oppression have been recognised across and between different groups of people in different places.

However, in acknowledging the undeniable value of the social model, it is easy to forget that it too is not without fault. In fighting to break down the myths of illness and dependence that have plagued disabled people's lives, Crow (1996) believes that recognition of the 'true' pain and inconvenience that an impairment can cause an individual has also been lost. At present society plays the dominant role in constructing disability, but the role of different physical and mental impairments cannot be ignored by the social model if it is to continue to be valued and respected. The heterogeneity of the different mind and body characteristics which can constitute impairments must not be forgotten, alongside recognition of the different people who embody these (for example, in

4

terms of gender, race and class: Crow 1996). It is not only over the lack of recognition of 'difference' that the social model has been criticised, but also that this approach, and the disability movement more generally, has been dominated by white, heterosexual, middle-class men (Morris 1992, 1993; Begum 1992). Such debates point to the difficulties involved in creating a critical and yet empowering 'disability studies' discipline, something which has perhaps been further compounded by contested relationships with other approaches and literatures.

It is useful to note here that there is a distinctive literature emanating from medical sociology which deals with chronic illness. From the early (heavily criticised) works of Parsons (1951) on the universal 'sick role' (a temporary social location for the 'sick', who are excused 'normal' responsibilities and social expectations: Barnes and Mercer 1996), to recent essays concerning the diverse lived experience of chronic illness and the varied ethical, social and political responses to such conditions of 'being in the world' (Toombs et al. 1995), there are a wealth of studies in this field. Deliberate attempts to bring disability literatures and chronic illness studies together are very recent, and fraught with problems:

> The portrayal of chronic illness by medical sociologists has also been criticised by disabled people as all too often a one dimensional catalogue of negative consequences and meanings – the stigma, 'loss of self' and dependence . . . Studies which suggest a more diverse experience, or which report a positive sense of self and creative involvement in the lives of disabled people are far less in evidence.
>
> (Barnes and Mercer 1996: 5)

Such a disjuncture between these literatures is slowly becoming less obvious, especially through the recent attempts to forge a useful dialogue. The interpretative focus of medical sociologists on the ill individual and the self can also have implications for the geographical analysis of impairment and disability. As Toombs et al. (1995: x) argue, for the person with chronic illness, 'there is a dialectical relationship between the individual and society – a constant reciprocity between subjective experience and the intersubjective milieu of everyday life'. The more structural and materialist foci of some disability studies (Gleeson 1996), where the 'structural' realm is much more than just intersubjective interactions, but rather the whole fabric of transforming capitalist relations in a globalising world, can, we would argue, be usefully combined with an interpretative focus on ill and impaired experience. By bringing such concerns together, the different spatial scales of the body, home, region, national and global (Smith 1993), which are necessarily involved in the analysis of impairment, illness and disability, can all be investigated.

Human geography and disability

Although it has been noted that geographers have for long been negligent in considering disability (instead tending to map diagnosed bodies in urban space: Giggs 1973; Mayer 1981), there is now a burgeoning literature within the discipline of human geography. In recent years geographers have increasingly engaged with both the theoretical debates surrounding 'different', ill or impaired bodies (terms which are discussed in more depth below) and the lived realities of ill/impaired experience (although primarily within a Western context: see Cormode 1997). The general rubric under which these issues have been considered is the 'geography of disability', a broad term capturing a wide range of interests, theoretical orientations and empirical studies. As Park *et al.* (1998) note, geographical research into disability and illness issues has fallen largely into two camps: broadly positivistic and behaviouralist (for example, Mayer 1981; Lovett and Gatrell 1988; Golledge *et al.* 1979, 1991; Golledge 1993, 1996), and post-positivistic work which has encompassed varied perspectives from social theory (for example, Dorn and Laws 1994; Hahn 1986, 1989; Imrie 1996; Laws 1994; Gleeson 1996). Crudely, we could map these two orientations on to the medical and social models of disability respectively. It is the latter direction of geographical enquiry which is of most interest to us here.

As geographers have become aware of the theoretical and empirical concerns of cognate disciplines such as cultural studies, anthropology, sociology and disability studies, so they have increasingly engaged with their literatures and become open to investigating spatial properties of varied social and cultural phenomenon. This diversity has prompted a deep engagement with key aspects of social and cultural life, such as the meanings and relations of gender, ethnicity and class (among many other concerns). The explosion of geographers' engagements with social theory has led to the discipline re-examining its key conceptual foundations of space and place, and in the process expanding the kinds of 'geographies' which can be discussed and theorised: geographies which range from the global to the nation-state to the living room to the body to the imagination. It is within such a sweep of perspectives that the 'geography of disability' has been revisioned, and it is also where this collection can be situated.

Even so, the 'geography of disability' is a label which hides much conceptual divergence in approaches to different mind and body states, and it is recognised that much recent work has attempted to specify exactly what it is here that geographers are investigating. For example, geographers are becoming more aware of the multiple critiques which can be made of 'ableism' and 'ableist geographies' (Chouinard 1997; Cormode 1997; Imrie 1996; Gleeson 1996, 1999), alongside the many trajectories of mind/body knowledges and how they are articulated and understood as 'disabled, impaired or ill' within both material and cultural landscapes. By using the work of Chouinard, we can usefully clarify these concerns. First, we can identify 'ableism' within geography, a concept which helps us understand the ways in which the discipline has ignored the bodily and social differences which shape many different human geographies. As Chouinard (1997: 380) outlines: 'ableism refers to ideas, practices, institutions and social relations that presume ablebodiedness, and by so doing, construct persons with disabilities as marginalised, oppressed, and largely invisible "others"'. Such ableism has arguably characterised the discipline of geography, which means that 'in geographic literature . . . able-body status continues to be assumed, helping to render disabling differences and their socio-spatial outcomes relatively invisible' (Chouinard 1997: 382). Second, we can identify 'ableist geographies' which are 'out there', in the world: 'Ableist geographies refer to lived environments which incorporate and perpetuate physical and social barriers to the participation of disabled persons in everyday life' (Chouinard 1997: 380). Ableism and ableist geographies are linked, in the sense that one informs the other, and with both arguably being perpetuated by a lack of critical engagement with such concepts by academics such as human geographers. It is also the case that the concept of 'ableism' may be more useful than the social model of disability, which, as noted previously, lacks attention to different mind and body states — something which critiques of ableism acknowledge.

With this basic premise in mind, we want to move forward to consider some important issues and influences which frame this particular collection, rather than to repeat the general overviews of geographical work on disability, the body and difference which have appeared within recent writings (see particularly Imrie 1996; Dear *et al.* 1997; Dorn 1994; Park *et al.* 1998).

New approaches

Wars over words: introducing impairment and illness

In adopting social theory perspectives, and in thinking through aspects of ableism, disability and difference, geographers have been forced to question their terminologies – and hence the meanings that are constructed within writing about geography and disability. Arguments between geographers about language have been revealing, as we can identify writers who more or less uncritically use particular terms and 'expert' categories (such as 'disabled', 'ill', 'normal', 'retarded', 'distorted') in reference to mind and body differences, and then we can identify other writers who insist that terms such as 'difference', 'ableism' and 'impairment' are more appropriate (Gleeson 1996; Golledge 1993, 1996; Imrie 1996; Parr 1997a) when thinking of bodies and minds which have particular physiological and psychological characteristics. Partly in recognition of these debates, and of the need to recognise lived experiences of mind and body differences as well as social constructions of such states, this introduction turns from an explicit discussion of the 'geography of disability' to consider the many geographies of illness and impairment which might be considered by a discipline sensitive to ableism and disablement (and hence the title of the book).

Using terms such as 'illness' and 'impairment' alongside 'disability' is not unproblematic, but what we want to signal here is that mind and body differences can be felt and conceived of in the following ways (some which overlap): as different states-of-being on a continuum of human mind/body characteristics; as a biomedical categorisation (commonly referring to an 'illness'); as a functional limitation (referring to an 'impairment'); or as a disability (referring to experiences of inequality due to physical and social barriers within society). The terms we have adopted here cross-cut and also differ from definitions offered by organisations such as the WHO, as well as by disability theorists and activists, but what we want to signal is that these overlapping understandings are a starting point which serve to link people and their experiences. Defining illness, impairment and disability alongside each other is difficult, and has political implications. We do not wish simply to collapse different experiences, contexts and minds and bodies onto each other to produce homogeneity, but we do want to highlight that different mind and body characteristics have perhaps been artificially separated by fixed understandings of what constitutes illness, physical impairment and mental healthiness.

An immediate qualification should perhaps be added with regard to 'illness', which can be both transitory and chronic, each having different implications for the individual. Transitory illness such as flu can have a large impact on an individual with physical impairments, more so than perhaps for someone without such bodily characteristics. However, such transitory episodes affect all people's lives, reminding us that the perfect, healthy body is indeed a myth. Chronic illness (a permanent and ongoing bio-physical or psychological condition which normally involves therapeutic interventions) is given most attention by medical sociologists who have made distinctions between the illness condition and illness as a social state (Barnes and Mercer 1996). For our purposes, we can none the less think of impairment and illness as different (involving a range of possible physiological and psychological characteristics, commonly eliciting quite particular medical and social understandings and responses), but also as similar in that individuals who fall into such categories (by self or expert definition) may be linked by experiences of frustrating mind/body characteristics, similar disabling structures, social responses and embodied acts of resistance in wider social life.

Thus, by arguing for impairment and illness to be considered alongside each other, we are effectively calling for the social construction of such states-of-being to be interrogated, and for both the structural and the experiential connections between these 'categories' to be exposed. As Dear *et al.* (1997: 455) argue, 'the ability to create and sustain distinctions between that which is the same and that which is different is dependent upon our ability to impose boundaries, that is, *partition*'. To partition further a phenomenon such as developmental difference, illness, impairment and mental ill/health may be justified, partially on the basis that very different physiological and psychological experiences are of importance, and, similarly, because many different systems of understandings, categorisations and care surround them (Oliver 1996). However, by making distinctions between (say) visual impairment, deafness, multiple sclerosis, learning difficulties, depression and schizophrenia, experiences of the body and mind are separated and often acquire different social and bio-medical categorisations which hold different 'values' and consequences within wider society. While on the one hand we think it is appropriate to recognise and differentiate between specific body and mind differences (we would welcome the writing through of social geographies of such particularities: for example, Dyck 1995), and also the responses they incur or require (whether medical treatment, policy developments or political action: Oliver 1996), we also think that there are some dangers when considering a range of social and biomedical categorisations as totally divorced from each other and as different from other concerns about (and geographies of) ableism and disability.

In relation to the points above, Dear *et al.*'s (1997) contention is useful, in that by blurring definitions and categories such as 'disability' (for example, to incorporate substance abuse, HIV, psychological conditions, arthritis), the resulting complexity means that conventional understandings of such terms and the embodied experiences in question are open to reinterpretation. Taking this argument one step further, we reject arguments that physical and sensory impairment, illness and mental difference are *entirely* separate personal, conceptual and political concerns. By taking this standpoint we are arguing for a continuum upon which to view bodily pain and certain psychological characteristics or disruption: we all experience these states to greater or lesser degrees. To state this is not to deny the specific recognition, and in certain cases assistance, which some states-of-being require in material, medical and social ways, but rather it is to deny that such difference has to equal a series of fixed and othering boundaries by which people are clearly defined and geographies are narrowly understood. This view signals that human-geographical work should interrogate the unacknowledged imaginings of shared communities of ablebodied/ableminded people which results in an othering of people whose bodies and minds do not meet with a mythical 'norm' (ableism). In this way, concepts and studies of body and mind differences may be further explored in a similar way to (and in association with) work on gender, race and class: key concerns in post-positivist human geography.

Although the authors in this book have not necessarily used these terms in exactly the same ways as here, they do still represent the varied mind/body states written about in the collection. We want to widen the scope of a 'geography of disability' to consider all sorts of people, with all sorts of different mind and body characteristics – we are interested in multiple aspects of their lives: their pain, their everyday geographies, their struggles, their positions within capitalist wage–labour relationships, how they are discriminated against, how they are represented and represent themselves, how social and physical environments are designed and built to exclude particular minds/bodies, how histories of ableism, medical categorisation and surveillance can enlighten us, and how people collectively and individually resist embodied and social limitations that mind/body differences can bring (however temporarily). In writing this introductory chapter, the editors do not position themselves outside of such concerns: both of us have experienced (to varying degrees and in different times and spaces) a variety of biomedical categorisations and interventions, experiences of limiting pain and environmental exclusion. We are hence claiming that different states of being, illness, functional and sensory impairment and mental differences can be considered as important aspects of life experience for *all* people, and that 'disability' references the disempowering social constructs and ableist structures which inappropriately surround all of us in our embodied, psychological worlds. The emphasis on the mind *and* body is deliberate as, following feminist and post-structuralist critiques of Western dualistic knowledges, these entities should be considered together and not forced apart. As Longhurst (1994: 215) has argued, 'like other disciplines of Western knowledge, geography has been built on a conception of the mind and body as separate and acting on each other'. In compiling this collection we are deliberately discarding traditional categorisations which separate bodily and mental difference, by claiming that these features of embodied selves are connected and affect all of us at different life stages. The 'geography of disability' can now perhaps be conceived as a diverse group of writings and materialities which encompass many aspects of embodiment and social construction, only some of which are represented here.

New medical geographies

At this point it is important to note that the developments in the wider discipline of human geography, hinted at earlier, have also begun to be reflected in the various subdisciplines. The engagement between medical geography and social theory has prompted recent work which has noted the importance of teasing out intersections between the body, diagnosis, care, therapy and place (Dorn and Laws 1994; Gesler 1992; Kearns 1993, 1994). These developments have implications for and connections with writings about geography and disability, impairment and illness, although such connections have been slow in realisation: 'It is ironic that medical geography, which draws its *raison d'etre* from a profession that is preoccupied with exploring the differences between the normal and the abnormal body, is itself so resistant to treating the body as a problematical concept' (Dorn and Laws 1994: 109). Despite this resistance, recent works are increasingly engaging with the body as a site of medical inscription and resistance, but crucially often in relation to place (Brown 1995, 1997a, 1997b; Dyck 1995; Kearns 1991; Kearns and Gesler, 1998; Wilton 1996). This critical engagement between mind/bodies and places is arguably a key concern of what could be termed 'new medical geographies'. In more 'traditional' medical geography (actually quite a diverse subdiscipline, not easily characterised: see Curtis and Taket 1996), the conceptualisation of what Kearns and Joseph (1993) have called an unproblematised 'geometric space' (concerned with distance and location) is being increasingly critiqued, with geographers noting that space cannot adequately be conceived of as a mere blank surface on which uncritically to map medical and 'deviant' subjects (Parr and Philo 1996). As part of this politicisation of space, the concept of place has assumed more theoretical significance, although for some this still means a lack of attention towards other spatialities such as that of the body (an important site of social relations and medical inscription and resistance): 'We believe that by continuing to ignore the social construction of the body and the struggles of new social movements, medical geographers will fail to take advantage of the lessons of social theory' (Dorn and Laws 1994: 107). The retheorisation of place in medical geography as a complex material, sociological, experiential and philosophical phenomena is crucial to thinking through how the local is involved in the making of and experience of different mind and body states (through place-based understandings of health, illness and the body, as well as appreciating the wider spaces of more structural contexts and responses to such phenomena: Kearns 1991; Kearns and Joseph 1993). However, as Dorn and Laws point out, this should be balanced with attention to powerful Western (and other) discourses about the body, as well as to the materiality of the body. We would also add to this that attention should be paid to the mind (and its potential for inscription with powerful Western discourse as well as the potential for partial understanding of the geographies of (un)consciousness: see Pile 1996).

Such general concerns within medical–geographical literature have been substantiated within particular works which have concentrated on intersections of place, bodies, identities and everyday geographies. It is within such works that links to geographical literatures on disability can be drawn. Dyck (1995) is one author who we will briefly mention, since she has broadly situated her writing as contributing to the recent conceptual directions in geographies of health (the 'new medical geography'). By focusing on the everyday lives of women with multiple sclerosis, and by tracing the physical, social and economic contexts which surround such embodied experience, Dyck achieves a geographically sensitive account of the multiple interconnections and negotiations which exist at and between the body and place:

11

close attention to the body in material context . . . provides the potential for exploring the involvement of dominant discourses and power
relations in the social construction of ideas about the body and identities, including that of the disabled body and the implications for the
experience of place, an area which is beginning to be of interest to
medical geographers.

(Dyck 1995: 308)

By further dismantling concepts such as 'place' into component entities such as
the neighbourhood, the street and the home, Dyck manages to 'map' material
spaces which are then *simultaneously* experienced and negotiated physically and
socially (through the mind and body). Dyck achieves a subtle series of insights
into the everyday geographies of women with MS as she pays attention to several
spatial 'layers' which contextualise these women's everyday lives (from residential relocation opportunities to restructuring home space and domestic
relationships). It is the *frission* between these accounts of everyday bodily spaces,
the formation of identity, the context of places and discourses of biomedicine
and embodied resistance which connects such work on the geographies of health
to literatures about impairment and disability, a connection which, until very
recently, has not been made all that effectively.

Mind and body: bringing mental differences into the picture

As part and parcel of the above concerns, and reflecting our general arguments
for the mind and body to be considered together, we have explicitly included
several chapters concerning the geographies of people with mental differences
(Milligan, Park and Radford, Parr). As already indicated, this is to emphasise
that the 'geography of disability' does not have to be closed around narrow physical definitions of what it is to be impaired and disabled, but rather it can –
perhaps should – embrace the multitude of embodied and behavioural characteristics which are seen as socially stigmatising and amenable to medical
categorisation and treatments (inscriptions which are resisted in many different
ways). The geographies of mental difference written about here are concerned
with historical understandings and spatialities of 'mental illness and deficiency',
as well as offering critiques of more recent re-locations of such categories and
peoples within mainstream social spaces (the move from institutionalisation to
deinstitutionalisation). Here the *embodied* experiences of diagnosis and medicalisation, and then of locations of care and support, are all addressed:

It is across and through this mish mash of sites – the hospital, the day
centre, the doctor's surgery, the drop-in clinic, the group home, the

12

night shelter, the soup kitchen – that mentally distressed people encounter a proliferation of discourses (and also concrete practices) which influence their identities in different ways, if only partially and if on occasion only because the individuals react against what they are hearing and experiencing.

(Parr and Philo 1995: 211)

There are parallels to be drawn between the experiences of deinstitutionalisation and mental health problems, which require a negotiation of the self (body and mind) at different spatial 'scales', and the women in Dyck's (1995) study. Both situations (experiences of mental health problems and experiences of multiple sclerosis) often involve complex negotiations of the ability to organise and to live out everyday geographies in home space; interactions with medical and social agencies; experiences of medication; implications for social networks in local neighbourhoods; and impacts on employment and housing opportunities (to name but a few possibilities). Experiences of mental health problems and multiple sclerosis both have corporeal and psychic dimensions: each having social and material implications that can be directly compared.

In order to expand on these arguments, it is important to bring a useful definition into play, one which connects with the concerns of this introduction and also the title of the book. 'Mind and body spaces' can be defined in a multitude of different ways depending on whether one is drawing on (say) humanistic, psychoanalytic or post-structuralist perspectives. Rather than become embroiled in different theoretical interpretations, however, we want to offer our own definitions of these terms so as to make clear what we understand to be the implications of considering both the mind and the body together. The term 'body space' has already been used by geographers in a specific context, referring to the corporeal terms by which gender and sexuality can be discursively 'placed' (Duncan 1996). In a similar vein, but in a more literal sense, we use the term 'body space' to refer to the physical, biochemical spaces of the body itself, and also to the immediate envelope of space which the body occupies in moving around and 'doing things'. In addition 'body space' does include a simultaneous set of discursive meanings, which are temporarily, culturally and geographically situated, and which serve to frame and to inscribe the body and the 'bodily possibilities' of any given situation. Indeed, there is no one universal body, but rather an always situated physical presence and actions which both reflect interpretations of the social world and are themselves interpreted in a multitude of ways. To use the term 'mind space' is even more problematic, and could lead us into discussions of spatial cognition and environmental psychology (Golledge and Timmermans 1990), but here we argue that at one level it must refer to the internal geographical imaginations of the (un)conscious, where identity and the self are constituted. However, we find ourselves

13

unable to offer a definition of 'mind space' without immediately referring back to the body given that 'feelings, impulses and thoughts are somewhere in the flesh' (Pile 1996: 87). Hence, we see the mind and body as connected, not reducible to either, but rather recursively constituted. By referring to mind and body spaces, we are thereby referring to the mutual importance and interrelationship of physicality and emotion, of the corporeal and the imaginative, and of the bodily and of identity. We are arguing, in a relatively straightforward manner, that the mind and body are not separate, but rather fused in complex physiological, psychological and sociological ways (which is why mental health problems and multiple sclerosis, for instance, cannot be seen as entirely separate concerns). The varied ways in which the mind and body are discursively constituted (together and separately) means that these relations are all too commonly disrupted, ignored and interpreted differently in different places. Although this approach to mind and body space is not without problems, we think it to be a useful one with which to introduce this collection, and to contextualise the inclusion therein of chapters tackling mental differences and mental health.

Taken as a whole, this collection also explores how notions of 'normality' are spatially and temporally specific. Central to constructions of normality and to productions of ableist spaces has been the separating out of corporeal and mental differences. This 'separating out' is seen as the outworking of dualistic understandings of the self and the other, the 'normal' and the 'abnormal', the productive and unproductive, the 'sane' and the 'insane', the attractive and the disfigured. The basis (both social and conceptual) of these dualisms is problematic, and involves diverse social constructions, including the attribution of value in capitalist labour processes, the influence of a normalising Western biomedicine, processes of 'abjection', and the manufacture of an 'ableist gaze' (Dear *et al.* 1997; Sibley 1995). While these dualisms and processes apply to both mind and body differences, many constructions render mental difference as somehow less desirable and acceptable than physical impairment (see Dear *et al.* 1997, despite ways of seeing physical difference as 'ugly' and 'monstrous'). Perhaps representing an 'ultimate other' (because of the supposed lack of 'reason' constructed as a central characteristic of post-Enlightenment human being and life), people with mental differences have been systematically shut away from mainstream Western society (Philo 1987, 1992, 1997) and have, only recently, re-entered everyday geographies of rural and urban areas in the wake of deinstitutionalisation policies (Parr 1997b). Hierarchies of othering are arguably based upon dynamics of anxiety, stereotyping and feelings about proximities (Dear *et al.* 1997). As we have indicated, the dynamics which intersect and produce *disabling* impairments, states-of-being and illness are varied and temporally and spatially specific, but, by considering mental differences alongside physical impairment and illness, we can see similar problems of social stigma, medicalisation and stereotyping. This perhaps provides a starting point towards dismantling such conceptual systems of othering. The three chapters mentioned above will look critically at socio-spatial separation and reintegration for people who have been categorised as 'mentally ill' and 'mentally deficient', but will also pay attention to the development of strategies of coping and resistance which come with the occupation of everyday contemporary social and medical spaces.

This collection

No one theoretical approach characterises this collection, which is eclectic and diverse in terms of conceptual and substantive materials. By differentiating between peoples and bodies, this book does not deal with an homogenous category of 'the disabled'. Instead, the various chapters acknowledge in different ways the varied biological and cultural contexts which surround and constitute the body and mind. All the chapters effectively seek to deny the reductionism which sees the mind/body as a generic or neutral core, and instead they emphasise a contextualism, a situatedness of understanding different mind/bodies in different places through multiple encounters. By providing both historical case studies and contemporary critiques of different geographies, this aim is realised by attempting to capture something of the diversity of ways in which mind and body differences have been understood and represented, are lived through, and have been spatialised and silenced (by an exclusionary society and discipline).

Imrie (1996: 48) argues, as he critiques constructionist, materialist and medical approaches to disability, that 'the different theorisations of disability, are, in their own ways, reductionist and unable to do justice to the multidimensionality of disablement'. Achieving a multidimensional view of 'disability' is not easy, and is undoubtedly not achieved by any of the chapters when considered alone, and probably requires more research than this collection offers. However, when viewed together these diverse writings tell us something about the multidimensionality that Imrie requests. As Birkenback notes:

> It strains one's imaging powers to try and locate disablement in a relationship between a medical and functional problem and the social responses to it, as the concept of disability requires.
>
> (Birkenback 1993: 178, cited in Imrie 1996: 48)

Although we are effectively advocating abandoning the term 'disability' as a catch-all category, these conceptual difficulties are ones which we feel this collection, taken as a whole, is attempting to address.

These theoretical concerns, then, are substantiated in the book by different authors in a variety of ways. Three chapters begin the collection by looking critically at *particular* histories of what we might call 'geometries of difference'.

Here the built environment and designs of urban spaces are critiqued as ableist through the writings of Le Corbusier, the famous Parisian architect. Imrie's work on Le Corbusier reveals how designs of the modern city contain powerful and often unacknowledged geographical visions of urban space and urban bodies. What is surprising here is that, despite the obsession with order, rationality and bodily perfection displayed in Le Corbusier's designs and writings, the messy, lived body continues to reassert itself within both the material spaces, and the conceptual thinking of the architect. This paper provides an assessment of Le Corbusier's conceptions of (architectural) space and suggests that these are premised upon a problematic decontextualised ideal of the body which is at the core of disabled people's oppression within the built environment.

Similarly, in Chapter 3, Dorn writes of another 'architect' of geographies and bodies through a critique of the works of Dr Daniel Drake, a medic who also had a powerful vision of the disordered and immoral bodies of alcohol drinkers. In a chapter which does much to highlight the social constructions of bodily deviance, Dorn highlights the material and moral geographies of the nineteenth-century American mid-West, showing effectively how an expanded vision of disability (to incorporate more than narrow views of physical impairment) and medical geography can highlight the myth and constructions of the perfect body. By focusing on geographies of temperance and intemperance, Dorn seeks to depict the particular body and gender politics that contextualised Drake's medical–geographic vision and to demonstrate how heavy drinking was constructed as a social disability in early nineteenth-century United States.

In Chapter 4, further 'geometries of difference' are introduced by Carter Park and Radford, who provide a fascinating account of the institutional segregation of people labelled as 'mentally deficient'. By empirically investigating the rhetorics surrounding 'intellectual disability' and place, these authors draw on materials from England, the United States and Canada. Carter Park and Radford use a four-pronged framework to explore the overlapping institutional discourses, treatment practices, geodemographic variables and social constructs surrounding such placings of intellectual difference. To understand such interrelationships, the rhetorics of cost, professional expertise and sexuality are all addressed with reference to the nation, morality, race and the body. In so doing this chapter links with those of Dorn and Imrie as a useful historical account of the construction of 'normality' and points to the varied processes involved in the identification of 'deviant' difference.

Moving from historical landscapes and accounts to more contemporary aspects of social and material life, Gleeson in Chapter 5 highlights arguments concerning technological determinism which, he argues, has characterised much of Western thinking about disability. Through a materialist framework, Gleeson suggests that the emancipatory potential of individual body-technologies, and also the critiques of the 'thoughtless designs' of built environments, ignore more fundamental

16

sources of division in contemporary society; namely capitalist wage–labour regimes. Such a materialistic interpretation of the deviance of the impaired and ill mind/body is balanced by two chapters which pay attention to capitalist wage–labour relationships, but from more interpretative and experiential positions. In Chapter 6, Dyck provides an insightful enquiry into the experiences of women with multiple sclerosis as they negotiate their position in the labour force. By using post-structuralist and feminist reference points, Dyck problematises the body and the workplace as sites of identity formation. An important point is made by Dyck as she argues that bodies which are in states of ongoing transformation and definition present challenges to the women concerned, but also present challenges about how such corporeal experience can be adequately conceptualised. Her discussion meets this challenge as she considers the intersections between the corporeal body, subjectivity and biomedical inscription and how these 'knowledges' are then negotiated and enacted within the micropolitics and social practices of the workplace.

Hall's Chapter 7 dovetails with these concerns as he presents an overview of issues and literatures related to disability and employment in the UK. Hall's key contribution here is his characterisation of how economic geographers have gone some way in understanding the social and bodily 'performances' required and regulated through work spaces in late twentieth-century capitalism. By intertwining these framing thoughts with a case study example, Hall seeks to understand how increasing demands in the service industry require bodies to be fluid and multifaceted between work roles and work spaces. Hall argues that, for people who experience impairment or illness, their bodily relationships to this 'fluidity' is problematic and a mesh of personal decision-making and employment restructuring serves to refigure employment opportunities and possibilities.

The ways in which people with impairments and illness negotiate these aspects of everyday life are often related to experiences of medical inscription alongside physical and psychological changes. Transformation in terms of the mind and body can take many forms, and does not always have to be conceived of negatively, as always limiting and painful. In Chapter 8, Moss discusses her own experiences of medical inscription as she was eventually diagnosed with ME. Moss frames these personal and empirical details by discussing the uses, forms and possibilities of the autobiographical as part and parcel of research projects in human geography. We take this chapter as a beginning point from which arises the potential to weave auto/biographies (of the researcher and participants in research) in and through positioned research concerning experiences of illness and impairment. Although we have reservations about the ability to always know the self in such depth (Rose 1997), Moss does highlight some interesting areas for further research, ones which point to positive outcomes and uses of such personal transformations.

17

In Chapter 9, Valentine also provides a fascinating account of the personal transformations of body and self by documenting an individual case study of a young
man (Paul) following an accident in which he was paralysed. Valentine explores
the concept of a hegemonic working class masculinity within which Paul and his
(un)conscious bodily performances may be located, what Valentine calls 'a shared
narrative of identity'. The chapter documents how this identity was threatened by
paralysis, and how Paul sought to re-establish a different set of bodily 'performances' through different geographies, ones which still affirmed, in different
ways, his masculinity. By conceptualising embodied identity, class and masculinity, Valentine achieves an insight into the ways in which the mind and body are
connected through transformation.

Parr, in Chapter 10, also discusses issues surrounding the body, self and identity by outlining the politics of taking medication as articulated by a group of
psychiatric service users. Parr's chapter documents the curious relationship
between the mind and body through the administration of psychiatric medication. Her case study of the 'user group' highlights how the meanings and negative
experiences of psychiatric medication are being resisted as the group advocate
the use of complementary medicines which effectively enable an empowering
union of body and mind and thus seek to disrupt conventional psychiatric medicine and discourse. Such a focus is unusual when discussing medical–institutional
geographies of mental health care and geographies of the body.

Staying with the theme of the corporeal and everyday life, in Chapter 11,
Butler looks at issues surrounding impairments, disability and sexuality. After
first discussing the significance of sexuality to self-identity, the chapter considers
the pressures of conformity and normalisation resulting in the homo/bisexual
and disabled person's need to 'pass' in everyday spaces. The restrictions which
impact upon an individual's freedom of expression are also examined in relation to the 'gay scene'. Here Butler considers the position of the disabled body
in gay culture, and the problems of access to the gay scene. Ableism within the
gay community, homophobia in the disabled community and discrimination within
'mainstream' society are all discussed in this new take on geographies of disability.

Widening the focus from the politics of the body as a site of medical inscription, limitation and transformation, we then have several chapters which offer
interpretations and case studies of the different ways in which society responds to
and represents impairment and illness. In Chapter 12, Milligan, focusing on the
findings of qualitative research conducted within a rural health authority region in
Scotland, considers factors that have contributed to the residential and everyday
geographies of a group of 'revolving door' patients who have experienced mental
health problems. Milligan discusses the processes of 'inclusion' and 'exclusion'
experienced by such individuals, and she argues that the varied influences of these
factors and other social and health care agencies have resulted in contradictory

locational forces in rural deinstitutionalised landscapes. It is arguably the case that, if the needs of individuals who experience impairment and illness were better understood, then the responses of specific care agencies – as well as society more generally – might be more sensitive to mind/body difference.

It is in this sense that Gathorne-Hardy makes her key contribution in Chapter 13. By focusing on questions of social justice, Gathorne-Hardy explores the processes involved in what she calls 'accommodating difference'. By presenting a case study of a housing project which designs and builds affordable housing for people with physical impairments, this chapter highlights the difficulties encountered in collaborative partnerships between individual households and 'progressive' community housing agencies. The author frames these empirical details with a discussion of the 'politics of difference' in relation to public policy, financial and institutional contexts. Her empowering conclusions argue that, rather than isolating disabled people's needs as 'special' and only to be locally addressed, more systematic 'accessible spaces' need to be created between all relevant scales of policy formation at national, regional and local scales in order radically to redirect political priorities so as to meet these housing needs.

Part of the problem which both Milligan and Gathorne-Hardy identify in relation to a mismatch between needs and provision is arguably related to the ways in which impairment, illness and difference have been represented within the popular imagination. In Chapter 14, Stables and Smith comment on issues of representation and particularly on the ways in which media representations of 'disabled people' carry laden meanings. In an innovative take on such issues, they concentrate upon child carers, the significant others who are implicated – but rarely discussed – in writings on disability and difference. The authors examine these new media narratives whose messages construct young carers as both tragic and heroic. Attention is also paid to how home space is constructed in particular ways by these representations. By examining the media's portrayal of young carers as 'poor victims' in home space, the authors conclude that such narratives are symptomatic of the social stigmatisation of people with impairments as 'bad' parents.

The final chapter in the collection is by Chouinard who comments on Canadian women's activism relating to issues surrounding disability. Chouinard focuses on several aspects of disabled women's political activism from the women's struggles for political voice to a discussion of the kinds of barriers which the women have encountered within both disability rights and women's movements when creating a body politics of resistance and transformation. Chouinard also notes the critical role of state restructuring in constraining and shaping disabled women's activism in a Canadian context. The chapter concludes with some reflections on the initiatives and alliances which exist for disabled women at local, national and global scales, and as Chouinard speculates, this has implications not only for empowering activism, but also for empowering analysis.

Overall, then, the book offers critiques of the diverse ways in which disability, illness and impairment can be interpreted. These critiques cover a variety of spatial scales and different aspects of social, cultural and material existence. From the social valuing of capitalist labour regimes to representations of child carers, from histories of constructed and moralised difference to the personal geographies of the mind/body, this collection effectively differentiates 'the geography of disability' by considering the intersections of ableist thinking, disabling structures and embodied experiences in private and public space. In particular, this book differentiates 'the built environment' (a key category of 'the geography of disability') into other categories of analysis such as the institution, the home, the workplace, public space and mind/body space, so as to think through the differences and similarities which face different people as they negotiate and create such geographies.

Conclusion

In a paper on 'Geography and the disabled . . .', Golledge makes the point that disabled people occupy what he terms 'transformed' and 'distorted' spaces (Golledge 1993: 64). Gleeson has critiqued this notion as effectively arguing for an ontologically distinct experience of social space (Gleeson 1996: 389), and for producing 'worlds of disability [which] . . . are seen to have a primary pathological genesis, located within the deficiencies of the disabled body rather than in social phenomena' (Gleeson 1996: 390). While others have also pointed criticism at Golledge's theoretical orientations, terminologies, lack of inclusion of individual, subjective viewpoints and failure to acknowledge wider social structures of oppression (Butler 1994; Imrie 1996; Parr 1997a), it is clear that he does have a concern about the immediate, real body spaces of disabled people. Indeed, in a more recent article Golledge (1997) writes a fascinating and revealing essay about the personal effects and impacts of vision loss, and asks geographers to focus more upon the personal – the literal bodily, organisational and emotional changes – which accompany such impairment. In arguing for this, however, he positions a more social vision of impairment and ableism (often articulated through the use of social theory) as a useless talking shop for political correctness: 'what has to be overcome is the debilitating effects of the problems of physical functioning. This depends on the will of the individual and the attitudes of friends and associates. It has little to do with political correctness or social theory or the politics of empowerment' (Golledge 1997: 409). While we are suspicious of this envisioning of a 'heroic' struggle against the individual broken body, it is not our intention to sideline Golledge's views as unimportant or unconnected to the aims of this book. In fact, Golledge might even be pleased with the attention of some chapters in this book to the bodily, the individual and aspects of personal struggle and organisation.

However, contrary to Golledge's dismissal of the importance of wider spatial scales and contexts of impairment (both physical and mental), we acknowledge that social, cultural, economic, political and medical frameworks do have very real impacts and effects upon the different, impaired and ill body/mind, including how it is represented and perceived. Not only are social attitudes important in defining what is and is not accepted as 'normal' (a question which Golledge does not see as overly problematic except to argue that 'normal' functioning does not happen for the 'disabled' person), but such boundary markers fuel and inform exclusionary actions from workplaces, night clubs, libraries, sex shops and myriad everyday spaces. Golledge may be annoyed with social and cultural geography for pointing out the ableism in social attitudes and actions, and he might wish that more could be written which would practically help disabled people in terms of spatial modelling and navigation technologies. Yet, if he had not been employed within (what seems to have been) a well-resourced and sympathetic work environment, he might be more sympathetic to writings which highlight the enormous difficulties faced by many people as a result of their socially constructed 'abnormality' in everyday spaces of social life such as the workplace. None of the authors of this book, we are confident to say, would deny or dismiss the real, lived experience of changed/changing/painful/clumsy/immobile bodies, nor of slow, confused, distressing, disrupted states of mind, but it is the very interconnections between such body/mind states and wider social, cultural, economic, political and medical environments which characterise this collection and, we would argue, *should* interest and concern us as geographers.

Acknowledgements

We would like to thank Chris Philo for invaluable comments and advice and Sophie Bowlby and Robyn Longhurst for useful and supportive reviews of the book.

References

Barnes, C. (1991) *Disabled People in Britain and Discrimination*, London: Hurst and Company.

Barnes, C. and Mercer, G. (1996) *Exploring the Divide: Illness and Disability*, Leeds: The Disability Press.

Barton, L. (ed.) (1996) *Disability and Society: Emerging Issues and Insights*, London: Longman.

Begum, N. (1992) 'Disabled women and the feminist agenda', *Feminist Review* 40 (Spring): 70–83.

Birkenbach, J. (1993) *Physical Disability and Social Policy*, Toronto: University of Toronto Press.

Brisenden, S. (1986) 'Independent living and the medical model of disability', *Disability, Handicap and Society* 1, 2: 6–13.

Brown, M. (1995) 'Ironies of distance: an ongoing critique of the geographies of AIDS', *Environment and Planning D: Society and Space* 13: 159–83.

—— (1997a) 'The cultural saliency of radical democracy: moments from the AIDS quilt', *Ecumene* 4, 1: 27–45.

—— (1997b) 'Radical politics out of place? the curious case of ACT UP Vancouver', in S. Pile and K. Michael (eds) *Geographies of Resistance*, London: Routledge.

Butler, R. (1994) 'Geography and vision-impaired and blind populations', *Transactions of the Institute of British Geographers* 19: 366–8.

Butler, R. and Bowlby, S. (1997) 'Bodies and spaces: an exploration of disabled people's experiences of public space', *Environment and Planning D: Society and Space* 15: 411–33.

Chouinard, V. (1997) 'Making space for disabling differences: challenges ableist geographies', *Environment and Planning D: Society and Space* 15: 379–87.

Cormode, L. (1997) 'Emerging geographies of impairment and disability: an introduction', *Environment and Planning D: Society and Space* 15, 4: 387–90.

Crow, L. (1996) 'Including all of our lives: renewing the social model of disability', in J. Morris (ed.) *Encounters with Strangers: Feminism and Disability*, London: The Women's Press.

Curtis, S. and Taket, A. (1996) *Health and Societies: Changing Perspectives*, London: Edward Arnold.

Dear, M., Gaber, L., Takahashi, L. and Wilton, R. (1997) 'Seeing people differently: the socio-spatial construction of disability', *Environment and Planning D: Society and Space* 15: 455–80.

Dorn, M. (1994) 'Disability as spatial dissidence: a cultural geography of the stigmatised body', unpublished MA thesis, Pennsylvania State University.

—— (1998) 'Beyond nomadism: the travel narratives of a cripple', in H. J. Nast and S. Pile (eds) *Places Through the Body*, London: Routledge.

Dorn, M. and Laws, G. (1994) 'Social theory, body politics and medical geography', *Professional Geographer* 46, 1: 106–10.

Duncan, N. (1996) *Bodyspace: Destabilitizing Geographies of Gender and Sexuality*, New York: Routledge.

Dyck, I. (1995) 'Hidden geographies: the changing lifeworlds of women with multiple sclerosis', *Social Science and Medicine* 40, 3: 307–20.

Finklestein, V. (1980) *Attitudes and Disabled People: Issues for Discussion*, New York: World Rehabilitation Fund.

Foucault, M. (1967) *Madness and Civilisation: a History of Insanity in an Age of Reason*, London: Tavistock.

Gesler, W. M. (1992) 'Therapeutic landscapes, medical issues in the light of the new cultural geography', *Social Science and Medicine* 34, 7: 735–46.

Giggs, J. A. (1973) 'The distribution of schizophrenics in Nottingham', *Transactions of the Institute of British Geographers* 59: 55–76.

Gleeson, B. (1996) 'A geography for disabled people?', *Transactions of the Institute of British Geographers* 21, 2: 387–96.

—— (1999) *Geographies of Disability*, London: Routledge.

Golledge, R. G. (1993) 'Geography and the disabled: a survey with special reference to vision impaired and blind populations', *Transactions of the Institute of British Geographers* 18, 1: 63–85.

—— (1996) 'A response to Gleeson and Imrie', *Transactions of the Institute of British Geographers* 21, 2: 404–11.

—— (1997) 'On reassembling one's life: overcoming disability in the academic environment', *Environment and Planning D: Society and Space* 15: 391–409.

Golledge, R. G., Loomis, J. M., Flury, A. and Yang, X. L. (1991) 'Designing a personal guidance system to aid navigation without sight: progress on the GIS component', *International Journal of Geographical Information Systems* 5, 4: 373–95.

Golledge, R. G., Parnicky, J. J. and Rayner, J. N. (1979) 'An experimental design for assessing the spatial competence of mildly retarded populations', *Social Science and Medicine* 13D: 292–5.

Golledge, R. G. and Timmermans, H. (1990) 'Applications of behavioural research on spatial problems: I cognition', *Progress in Human Geography* 14: 57–99.

Hahn, H. (1986) 'Disability and the urban environment: a perspective on Los Angeles', *Environment and Planning D: Society and Space* 4: 273–88.

—— (1989) 'Disability and the reproduction of bodily images: the dynamics of human appearances', in J. Wolch and M. Dear (eds) *The Power of Geography: How Territory Shapes Social Life*, London: Unwin Hyman.

Imrie, R. F. (1996) *Disability and the City*, London: Paul Chapman.

Kearns, R. A. (1987) 'In the shadow of illness: a social geography of the chronically mentally disabled in Hamilton, Ontario', unpublished Ph.D. thesis, McMaster University.

—— (1991) 'The place of health in the health of the place: the case of the Hokianga special medical area', *Social Science and Medicine* 33, 4: 519–30.

—— (1993) 'Place and health: towards a reformed medical geography', *The Professional Geographer* 45: 139–47.

—— (1994) 'Putting health and healthcare into place: an invitation accepted and declined', *The Professional Geographer* 46, 1: 111–15.

Kearns, R. A. and Gesler, W. M. (eds) (1998) *Putting Health into Place: Landscape Identity and Wellbeing*, New York: Syracuse University Press.

Kearns, R. A. and Joseph, A. E. (1993) 'Space in its place: developing the link in medical geography', *Social Science and Medicine* 37, 6: 711–17.

Laws, G. (1994) 'Oppression, knowledge and the built environment', *Political Geography* 13, 1: 7–32.

Longhurst, R. (1994) 'The geography closest in – the body . . . the politics of pregnability', *Australian Geographical Studies* 32, 2: 21–223.

Lovett, A. A. and Gatrell, A. C. (1988) 'The geography of spina bifida in England and Wales', *Transactions of the Institute of British Geographers* 13: 288–302.

Mayer, D. (1981) 'Geographical clues about multiple sclerosis', *Annals of the Association of American Geographers* 71, 1: 28–39.

Morris, J. (1992) 'Personal and political: a feminist perspective on researching physical disability', *Disability, Handicap and Society* 7, 2: 157–66.

—— (1993) 'Feminism and disability', *Feminist Review* 43 (Spring): 57–70.

Oliver, M. (1984) 'The politics of disability', *Critical Social Policy* 4, 2: 21–32.

—— (1990) *The Politics of Disablement*, Hampshire: Macmillan.

—— (1996) 'Defining disability and impairment: issues at stake', in C. Barnes and G. Mercer (eds) *Exploring the Divide: Illness and Disability*, Leeds: The Disability Press.

Park, C. D., Radford, J., Vickers, M. H. (1998) 'Disability studies in human geography', *Progress in Human Geography* 22, 2: 208–33.

Parr, H. (1997a) 'Naming names: brief thoughts on disability and geography', *Area* 29, 2: 173–6.

—— (1997b) 'Mental health, public space and the city: questions of individual and collective access', *Environment and Planning D: Society and Space*, 15: 435–54.

Parr, H. and Philo, C. (1995) 'Mapping mad identities', in S. Pile and N. Thrift (eds) *Mapping the Subject: Geographies of Cultural Transformation*, London: Routledge.

—— (1996) *A Forbidding Fortress of Locks, Bars and Padded Cells? The Locational History of Nottingham's Mental Health Care*, Historical Geography Research Series. No. 32.

Parsons, T. (1951) *The Social System*, New York: Free Press.

Philo, C. (1987) 'Fit localities for an asylum: the historical geography of the nineteenth-century 'mad-business' in England as viewed through the pages of the Asylum Journal', *Journal of Historical Geography* 13, 4: 398–415.

—— (1992) 'The space reserved for insanity: studies in the historical geography of the mad business in England and Wales', unpublished Ph.D. thesis, University of Cambridge.

—— (1997) 'Across the water: reviewing geographical studies of asylums and other mental health facilities', *Health and Place* 3, 2: 73–89.

Pile, S. (1996) *The Body and the City*, London: Routledge.

Rose, G. (1997) 'Situating knowledges: positionality, reflexivities and other tactics', *Progress in Human Geography* 21, 3: 305–20.

Shakespeare, T. (1993) 'Disabled people's self organisation: a new social movement?', *Disability, Handicap and Society* 8: 249–64.

Sibley, D. (1995) *Geographies of Exclusion*, London: Routledge.

Smith, N. (1993) 'Homeless/global: scaling places', in J. Bird, B. Curtis, T. Putnam, G. Robertson and L. Tickner (eds) *Mapping the Futures: Local Cultures, Global Change*, London: Routledge.

Swain, J., Finkelstein, V., French, S. and Oliver, M. (1994) (eds) *Disabling Barriers – Enabling Environments*, London: Sage.

Taylor, S. M., Elliot, S. and Kearns, R. (1989) 'The housing experience of chronically mentally disabled clients in Hamilton, Ontario', *The Canadian Geographer* 2: 146–55.

Toombs, S. K., Barnard, D. and Carson, R. A. (1995) *Chronic Illness: From Experience to Policy*, Indianapolis: Indiana University Press.

Turner, B. (1987) *Medical Power and Social Knowledge*, London: Sage.

UPIAS (1976) *Fundamental Principles of Disability*, London: Union of the Physically Impaired Against Segregation.

Wilton, R. D. (1996) 'Diminished worlds? the geography of everyday life with HIV/AIDS', *Health and Place* 2, 2: 69–83.

Zola, I. (1972) 'Medicine as an institution of social control', *Sociological Review* 20: 487–504.

2

THE BODY, DISABILITY AND LE CORBUSIER'S CONCEPTION OF THE RADIANT ENVIRONMENT

Rob Imrie

Introduction

Proportion is the commensuration of the various constituent parts with the whole. For no building can possess the attributes of composition . . . unless there exists that perfect conformation of parts which may be observed in a well formed human body.

(Vitruvius 1960)

One of the critical contexts for the perpetuation and reproduction of social inequalities is the built environment (Crowe 1995; Knox 1987; Laws 1994a, 1994b). For disabled people in particular, the built environment is often encountered as a series of hostile, exclusive and oppressive spaces. Examples abound of discriminatory architectural design, including steps into shops and public buildings, inaccessible transport, and the absence of colour coding and induction loops. Indeed, most housing in the United Kingdom is not wheelchair accessible, yet for the House Builders Federation (1995) this is barely an issue.[1] As they state, 'if a disabled person visits a homeowner, it is to be expected that they can be assisted over the threshold' (HBF 1995: 1). Moreover, in the 1997 British general election, 75 per cent of polling offices were inaccessible to people in wheelchairs, while few contained the technical aids to permit visually impaired and/or blind people to mark their vote on the polling papers. In Lefevre's (1968) terms, such representations of space project the dominant values of specific body-types, that is, the 'able-bodied', or bodies characterised by a 'statically balanced symmetrical figure with well defined limbs and muscles' (McAnulty 1992: 181).

Ableist bodily conceptions underpin architectural discourses and practices, and there is evidence to suggest that the specific mobility and/or access needs of disabled people rarely feature in the theories and practices of designers or archi-tects (Davies and Lifchez 1987; Hayden 1981; Weisman 1992).[2] In this sense, one of the sources and sites of disabled people's marginalisation and oppression in society relates to architects and architecture (also see Dickens 1980; Imrie 1996; Knesl 1984; Knox 1987). In particular, architectural conceptions of the body are premised upon abstract theories of the self, or what Lester (1997: 481) refers to as 'a largely disembodied self which is held to be outside of time, space, outside of culture and gender'. Yet, as Lester (1997) comments, the presump-tion of a disembodied self is impossible and what generally has been presented through the context of architecture, art, and other mediums is less a body in a neutered state but one infused with (male) gender, class, and the embodiment of health and normality (see, in particular, Probyn 1993).

In this chapter, I develop the proposition that modern architectural concep-tions of architectural form and the built environment are simultaneously ableist and disablist by ignoring and/or denying the multiplicities of the human body. I suggest that such conceptions are premised upon a decontextualised, disem-bodied, ideal of the body which is at the core of disabled people's oppression within the built environment. In pursuing such themes the chapter is divided into three. The first part is a brief discussion of architectural modernism and the emergence of what some have referred to as disembodied architecture (Colomina 1994; Gray 1929; Grosz 1994, 1995; Mumford 1968; Whiteman et al. 1992). In the second section, I relate such ideas to a consideration of the architectural theories and practices of one of the most influential architects of the twentieth century, Le Corbusier. How, for example, did Le Corbusier conceive of the human body and of its possible multiplicities and how did such conceptions of body/mind inform, if at all, his architecture and approaches to urbanism and urban planning? I conclude by discussing the possibilities for the development of non-ableist, embodied, architectural discourses and practices.

Modernity, technology and decontextualised/ disembodied architecture

The theories and practices of architects in Western society are underpinned by what Grosz (1995: 127) refers to as 'epistemic domains where the neutrality, transparency, and universality of the body is all assumed'. Such epistemic concep-tions are connected to modernist values which have been highly influential in twentieth-century architecture (Le Corbusier 1925a, 1925b, 1967; Sullivan 1947; also, see Ward 1993). For Sullivan (1947), for example, such values involved the search for a 'true normal type' and for universal laws of human habitation and behaviour. Underpinning this was the propagation of an engineering aesthetic based on the idea that pure, distilled, design could be produced which was fixed and absolute, singular, transcontextual, and grafted from the essence of the human being. Such essentialism has its roots in classical ideas where the body was conceived of in naturalistic terms or what Grosz (1995) refers to as the cause and motivation for the design of cities (also see Colomina 1994; Frampton 1980, 1991; Sennett 1994). Indeed, as McAnulty (1992: 182) notes, 'the unified body of the Vitruvian image was taken as the model for classical architecture' in that buildings were to replicate its order, harmony, and proportions.

As Plate 2.1 shows, the essentialist depiction of such bodies is of geometric proportion and symmetry, or, in this instance, of a body seemingly able-bodied, taut, upright, male, an image projected as self-evidently invariable, normal, vigorous and healthy. The body (to be built for) was conceived of, in classical terms, as being constituted prior to its projection into the world, comprising what McAnulty (1992: 182) refers to as a 'figural self sufficiency'. In this sense, the body was posited as a purity, prior to, and beyond, socialisation or culture (on this theme see, for example, Bordo 1995; Grosz 1995; Shilling 1995). Such bodies were either seen as being reducible to organic or technical and instrumental matter, that is, machine-like or, as Grosz (1995: 8) notes, 'merely physical, an object like any other'. Thus, such bodies are without sex or gender, or class or culture. They are, in Ann Hall's (1996) terms, objective entities to be dissected, manipulated, treated, and utilised as instruments and/or objects.[3]

Such conceptions, in informing the values and practices of architects, are also premised upon a body assumed to be an organic system of interrelated bits, pieces of matter alike in functioning and form. For architects, the body, as somehow inert, passive, and pliable, is a pre-given which permits its (geometric) proportions to define the possibilities of design and building form. The body, then, as pre-formed, fixed, and known, has led some to refer to the ideas and practices of architects as necessarily leading to the production of 'standard-fit' design, that is, decontextualised, one-dimensional, architecture (Colomina 1994; Grosz 1995; Tschumi 1996). Thus, as Gray (1929, in Nevins 1981: 71) commented, in referring to the rise of the abstractions of modern or avant-garde architecture:

> the avant-garde is intoxicated by the machine aesthetic . . . But the machine aesthetic is not everything . . . Their intense intellectualism wants to suppress everything which is marvellous in life . . . Their desire for rigid precision makes them neglect the beauty of all these forms . . . Their architecture is without soul.

For Gray (1929, in Nevins 1981), the machine aesthetic was premised on universalising the essence of the body by a denial of the (contextual and contextualised) differences in bodily experiences and form. As Ward (1993: 43) suggests, the

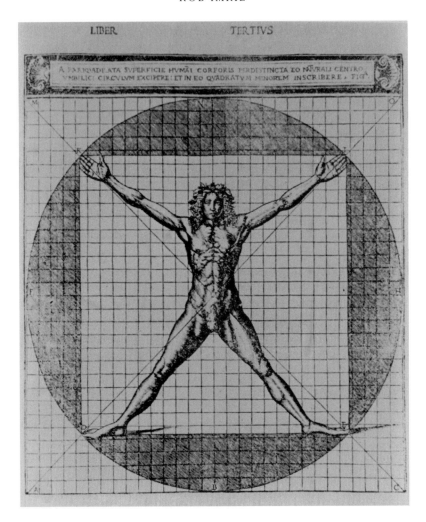

Plate 2.1 Vitruvian image of the body

rationality underpinning such meta-narratives 'erases differences, standardises experiences, drains the world of colour and texture, and precludes the richness and quality of life'. This, then, is a world which seeks to normalise.

Such conceptions of architecture and bodies are also problematical for conceiving of buildings, bodies, and environments as discrete, rather than constitutive, entities. Indeed, for Knox (1987: 355) architecture, buildings, and the wider built environment have been assigned the roles of independent variables 'explaining everything from people's perceptual acuity to their social networks'. In this sense, form is seen as shaping space and, in turn, space is conceived of as giving shape to social relations. Such determinism was underpinned by what Gray (1929) referred to as the vain arrogance of architects in their popularisation of the aesthestic, or form, over the (bodily) use of buildings, so conceiving of the idea that the architect as artist is instilling critical capacities into buildings. And, as Tschumi (1996) notes, bodies have been regarded as problematical in the architect's wider, critical, endeavours, in being seen in Platonic terms as impure, degenerate and for their potentially disruptive influence on aesthetic and/or design considerations (also see Frampton 1980; Ghiraldo 1991).

A slippage between the categories of mind/body, art/craft, architect/ craftsperson, purity/impurity, etc., is important in supporting and sustaining the position of architects as artists or purveyors of beauty and truth (see Figure 2.1). For Le Corbusier, and others, there was little doubt as to the elevated status of the architect over those without the requisite critical capacities (Sullivan 1947). As Le Corbusier (1925a: 137) stated: 'architecture is there, concerned with our home, our comfort, and our heart. Comfort and proportion. Reason and aesthetics. Machine and plastic form. Calm and beauty'. As Figure 2.1 suggests, the rationality of the avant-garde was sustained by the conception of the architect as the mindful and ethereal purveyor of good taste, or as those who were able to create the rational disposition of spaces beyond the contamination of the earthy impurities of society. Yet, as Caygill (1990: 261–2) notes, such claims, to represent the wider world, necessarily depend on abstracting 'from individual idiosyncrasies and differences in order to reduce them to complexes of universal human needs and rights'.

While such conceptions have been difficult to sustain in the wider social sciences and humanities, they still underpin many of the theories and practices of architects (see, for example, accounts and critiques by Jencks 1987; Knox 1987; Venturi 1966; Wolfe 1981). The abstract premises of modernism are, in

Mind	Body
Art	Craft
Artist	Craftsperson
Architect	Builder
Beauty	Attraction
Truth	Contingency
Facts	Values

Figure 2.1 Positioning architecture within Cartesian dualisms

Caygill's (1990) terms, particularly problematical for divorcing conceptions of building form (body) and design (mind) from its use. This, for Caygill (1990), is implicated in the production of insensitive, decontextual, design. For others, the limitations of the avant-garde are expressed through the materiality of designed body-spaces premised on conceptions of standard body sizes and shapes, that is, the body as objectification. However, for Merleau-Ponty, and other theorists, the body is not an object *per se* 'but it is the condition and context through which an embodied person is able to have a relation to objects' (Merleau-Ponty 1962, quoted in Grosz 1995: 5; also see Ghiraldo 1991; Probyn 1993). In this sense, the body constructs, and is constructed by, 'an interior, a psychical and a signifying viewpoint, a consciousness or perspective' (Grosz 1995: 8).

However, such critiques, and reformulations of body/mind, have had little effect on the writings and practices of most architects and there is no evidence to suggest that architectural schools teach trainee architects about the problems and limitations of decontextual conceptions of the body and architecture.[4] Moreover, there has been little written about the specificity of bodies in the ideas and practices of the more influential architects and architectural traditions (although, for notable exceptions, see Colomina 1994; Tschumi 1996; Whiteman *et al.* 1992). Little or nothing has been documented about whether or not architects and their practices are self-consciously sensitised to diverse conceptions of, for instance, disabled bodies. In seeking to redress, in part, this research lacuna, the rest of the chapter is a preliminary exploration of the writings of one of the most influential architects of the twentieth century, Le Corbusier. The objective is twofold. First, by referring to the earlier work of Le Corbusier, between 1924 and 1933, I provide a brief documentation of the ways in which he wrote about, and conceived of, the human body. Second, I relate Le Corbusier's conceptions of the human body to examples of his architecture. In particular, I refer to 'the Ville Radieuse' or Le Corbusier's conception of a radiant or ideal city to consider how conceptions of the body were drawn into the specificity of (his) design.

In documenting aspects of Le Corbusier's writings and architectural practices, two qualifications should be made. First, following Jencks (1987), the life of Le Corbusier was characterised by significant changes in perspectives and modes of thinking. By the second half of his career, from the early 1940s, Le Corbusier's conceptions of the interrelationships between body/mind and design had shifted to a degree that it is impossible to gauge any temporal unity and/or consistency to his thinking on such matters. Second, Le Corbusier's writings are idiosyncratic and obscure, seemingly half-formed, yet full of detail and contradiction. In this sense, this chapter is a small contribution to a wider research agenda in which much remains to be done to excavate the meanings, materialities, and processes of bodies in Le Corbusier's architectural spaces.

Spatial perfectibility, Le Corbusier and the radiant environment

The architect Le Corbusier was characterised by Mumford (1965: 34) as a 'crippled genius' who 'warped the work of a whole generation, giving it arbitrary directives, superficial slogans, and sterile goals'. Others have noted the one-dimensional conception of the body propagated by Le Corbusier, the portrayal of people as asexual, and his pre-occupation with the establishment of an 'able-bodied' standard in order to face what Le Corbusier characterised as the problem of perfection (Colomina 1994). Le Corbusier was influenced by modernism and the emergent avant-garde of the 1920s and with the search for what Tschumi (1996) refers to as the specificity of architecture. Such specificity, for Le Corbusier (1927: 220), was conceived of, in part, as an art of geometric volumes or, as he argued, 'there is nothing but pure forms in precise relationships'. Le Corbusier was influenced by classical design and by the desire to conceive of architecture as a 'process based on standards' (Le Corbusier 1980: 37). Such standards were, for Le Corbusier, based upon the 'truths and emotions of a superior mathematical order' (Le Corbusier 1927: 221).[5]

Le Corbusier rejected what he saw as the disequilibria of curved lines and jagged edges for what he characterised as 'the classical equilibrium of rectangles and pure volumes' (1980: 23). His inspiration was connected to the specificity of precision. Such precision, for Le Corbusier (1980: 18), was evident in classical architecture:

> the Parthenon, the Indian Temples, and the cathedrals were all built according to precise measures which constitutes a code, a coherent system: a system which proclaimed an essential unity.

Le Corbusier was, as Jencks (1987: 112) notes, 'propelled by a single vision of technology as a progressive force which, if guided by the right ideals, might reinstate a natural and harmonic order'. Underpinning such conceptions of architectural form and process was purism, a system of thinking which was concerned about 'the laws of natural selection which inevitably produces the pure forms of standardised objects' (Le Corbusier 1925a: 74). Purism was also obsessed with the typical or, as Curtis (1986: 50) argues, 'the purist thought that neither the human figure nor landscapes were relevant to their aims . . . they wished to portray familiar everyday objects and to raise them to the levels of symbols by extracting their most generalised features'. Thus, purism was important in leading Le Corbusier to a distillation of the body's essence, of a thing, and of conceiving the body in ideal-typical terms. Such conceptions are evident in Le Corbusier's writings and, as he notes:

the establishment of a standard involved evoking every practical and reasonable possibility and extracting from them a recognised type con-
formable to all functions with a maximum output and a minimum use of
means and workmanship and material, words, forms, colours, sounds.

(Le Corbusier 1980: 27)

The idea of a type is also evident in *La Peinture Moderne*, where Le Corbusier
(1925b: 83) develops the notion that 'mechanical evolution leads at once towards
the universal and the geometrical culminating in the slogan that man is a geometrical animal'. For Le Corbusier (1925b: 83) such geometric specificity could be
related to the search for the human type or a form of 'universal symbolism that
would be trans-historical' (also see Jencks 1987; King 1996). As Le Corbusier
(1925a: 72) suggested:

> to search for the human scale, for human function, is to define human
> needs. They are not very numerous; they are very similar for all
> mankind, since man has been made out of the same mould from the
> earliest times known to us . . . the whole machine is there, the struc-
> ture, the nervous system, the arterial system, and this applies to every
> single one of us exactly and without exception.

Such conceptions of the body were recurrent in the writings of Le Corbusier in
the 1920s and early 1930s. Throughout this period, Le Corbusier's conceptions
of the body were derived from the Newtonian idea of the body as a machine,
that is, in Sennett's (1994: 7) terms, 'a closed system, mechanical, with all of
its parts rigid and pre-given to interaction'. For Colomina (1994: 136), Le
Corbusier conceived of the body as a 'surrogate machine in an industrial age'.
Indeed, there are clues in Le Corbusier's writings which support Colomina's
contention, that is, of the body as analogous to a type reducible to specific,
interlocking, mechanical parts. Thus, as Le Corbusier (1925a: 76) intimated:

> If our spirits vary, our skeletons are alike, our muscles are in the same
> places and perform the same functions: dimensions and mechanism are
> thus fixed . . . human limb objects are in accord with our sense of
> harmony in that they are in accord with our bodies.

The fixity of the (physical) body for Le Corbusier reaffirmed the underlying
essentialist terms in which he wrote about the interconnections between bodies
and architecture. For instance, in referring to the human (bodily) organism as
a machine, Le Corbusier (1947: 22) conceived of biology as the determinant of
human need or, as he said: 'since all men have the same biological organisation,

they all have the same basic needs'. As Le Corbusier (1925a: 71) suggested, the specificity of the body was 'a man, a constant, the fixed point'. The body, for Le Corbusier, was, therefore, a biology and/or a physiology of parts which did not vary. For example, in a typical flourish, Le Corbusier (1925a: 33) comments about his conception of the standard body:

> the climates, the suns, the regimes, the races, everything is classified in terms of its relationship to man. A typical, standardised, normal man: two legs, two arms, a head. A man who perceives red, or blue, or yellow, or green.

In turn, his architecture was, so some argue, based on a world which seemingly denied the relevance of difference or the vitality of the knowing (individual) subject (see, for example, Frampton 1980; Colomina 1994; King 1996; Wigley 1992). In an interchange between Le Corbusier and an unnamed individual, Le Corbusier's views, on the potentially complex interrelationships between the body and architectural form, reinforced his belief in the possibilities of generic solutions to fit the standard or constant-type. Thus, as Le Corbusier (1925a: 72) commented:

> nevertheless, one of the big names of the 1925 Exhibition recently violently disagreed; with his heart set on multifold poetry, he proclaimed the need of each individual for something different claiming different circumstances in each case; the fat man, the thin man, the short, the long, the ruddy, the lymphatic, the violent . . . he sees the character of an individual as dictating his every act.

However, in dismissive terms, Le Corbusier (1925a: 72) noted: 'let us recognise the practical impossibility of this dream of an individual sentient object, in all its intimate multiplicity'. He reinforces this further on in the text when commenting that 'in all things that are in universal use, individual fantasy bows before human fact' (76). For Le Corbusier, then, the distillation of bodily essence was critical in developing systems of standards and measures which, in turn, were to be used in designing the built environment. In rejecting the individual sentient-object, Le Corbusier conceived of a world where the (standardised) measurements of the body would be critical in giving shape to the objects, decorations, and materials of everyday (human) use.

For Le Corbusier, everything external to the body is but an extension of the body, or what he termed human-limb objects. In deterministic terms, he conceives of a close interdependence between the body and the processes of conception and production of the wider material world of objects, fixtures, and decorations, a critical interdependence which he referred to as 'decorative art'. Thus, for Le Corbusier (1925a: 72):

decorative art is the mechanical system which surrounds us . . . an extension of our limbs; its elements, in fact, artificial limbs. Decorative art becomes orthopaedic, an activity that appeals to the imagination, to invention, to skill, but a craft analogous to the tailor; the client is a man, familiar to us all and precisely defined.

Such ideas were also extended by Le Corbusier to the design of fixtures and fittings within buildings, including furniture. As he (1967: 22) suggested:

furniture should be reduced to sectional compartments, built to hold objects used by human beings. They can all be made in a common scale with standard measurements.

For Le Corbusier, such statements were self-evident given the type-nature of the body. All human needs were conceived by Le Corbusier (1925a: 72) as similar or, as he noted:

These needs are type, that is to say they are the same for all of us . . . since nature is indifferent, inhuman (extra human), and inclement, we are all born natural and with insufficient armour.

Such views were exemplified in an exhibition of Le Corbusier's furniture for the Salon d'Automne in 1929 where he reaffirmed his belief in starting the design process from basic functional requirements which he saw as constant and universal. Thus, as Le Corbusier (1925a: 73) said about the design of the furniture: 'let us appreciate the wisdom (the anthropocentric mean) that established it'. For Le Corbusier, then, the design process was one of evolution and experimentation, of modelling the basic postures of the human body until the standard or object type had been derived.

Le Corbusier's conceptions of the body, as geometrically proportionate, and ideal-type, were, not surprisingly, interconnected to his architecture and to wider conceptions of urbanism and urban planning. Throughout the 1920s and early 1930s, his binding world view was, according to Curtis (1986: 60), 'linked to the definition of ideal types to save the industrial city from disaster'. Such views evoked naturalistic and anti-urban conceptions of cities as places of despoilation and degradation. Thus, for Le Corbusier (1967: 93), 'the city is swelling, the city is filling up . . . all the houses are on streets, the street is the basic organ of the city . . . the street becomes appalling, noisy, dusty, dangerous'. References of this type are littered throughout his writings, where medical metaphors abound in conjuring up the horrors of contemporary urbanism. In The Radiant City (1967), Le Corbusier presents disturbing images which, none the less, are caricatures. As he (1967: 94) stated:

Our cities are too old; they are crumbling away; they are uninhabit-
able; they are full of lurking disease; it is impossible to move around
in them anymore; traffic has reached its ceiling and the reign of speed
is leading to total immobility.

Overturning the socio-environmental degradation of the cities was, for Le
Corbusier, the object of the architect or, as he (1927: 16) said, 'the architect by
his arrangement of forms, realises an order which is a pure creation of the city'.
Such creations were seen as the embodiment of health, vigour, cleanliness and, in
particular, the youthful body. As Le Corbusier (1947: 115) commented:

Youth gathers round, ready to participate, eager for instruction, desiring
to bring its youthful energy to work which it understands so well because
it itself must soon feel the benefits.

The symbiosis between beauty, truth, environment, and health was, for Le
Corbusier, connected to a specific conception of providing for the 'good body' or
a body which, by implication, denigrated aged and disabled bodies. Architecture,
not revolution, was Le Corbusier's solution to his perception of urban degradation premised on 'youth indeed, and [which] rightly, claims to gather the flower
of a root planted with that intention' (Le Corbusier 1947: 119).

In particular, Le Corbusier's use of body metaphors reveals his search for
forms of spatial perfectibility premised on a pure body-type conceived of as the
youthful, normal or classical body. For instance, he (1967: 92) characterised the
emergent suburbs of the 1920s and 1930s as 'broken dislocated limbs' charac-
terising cities as having 'been torn apart and scattered in meaningless fragments
across the countryside'. In biblical tones, Le Corbusier (1967: 92) argued the
point in that 'suburban life is a despicable delusion entertained by a society
stricken with blindness'. Thus, for Le Corbusier, early twentieth-century
urbanism was analogous to the negativisms of twisted deformities, broken body
parts, even visual impairment. As he (1967: 92) stated:

The world is sick. A readjustment has become necessary. Readjustment?
No, that is too tame. It is the possibility of a great adventure that lies
before mankind: the building of a whole new world . . . because there
is no time to be lost.

For Le Corbusier, the readjustment was premised upon an ideal-type of the
'good environment', a conception which he termed the *Ville Radieuse* or the
Radiant City (for good accounts on the Radiant City, see Frampton 1980; Jencks
1987). The Radiant City was, as depicted in Plate 2.2, conceived by Le Corbusier

as a rational, planned, social order premised on a particular configuration of space achieved through the application of technology and mass production systems building. As Le Corbusier (1967: 94) argued:

> the problem is to create the Radiant City . . . the city of light that will dispel the miasmas of anxiety now darkening our lives . . . its growth is assured. It is the Radiant City. A gift to all of us from modern technology.

The planned disposition of space was a form of poetics, while the eradication of the messiness of the street and the restoration of clean air and open spaces were conceived of as pre-requisites for 'the rebirth of the human body' (Le Corbusier 1967: 7). The Radiant City was, therefore, an organism capable of restoring vitality to the human condition under assault from contemporary urbanism. As Le Corbusier (1967: 94) claimed:

> The Radiant City, inspired by the laws of the universe and by human law, is an attempt to guarantee the men of a machine civilisation all this world's best pleasures. For all men, in cities and in farms: sun in the house, sky through their windowpanes, trees to look at as soon as they step outside.'

However, the Radiant City was also premised on a specific and problematical conception of the (able) body, utilising biological analogies to identify interlocking (bodily) functions. As Le Corbusier (1947: 2) commented:

> a plan arranges organs in order, thus creating organism or organisms. BIOLOGY! The great new word in architecture and planning . . . these skyscrapers (of the city center) will contain the city's brains, the brains of the whole nation.

The Radiant City was divided into seemingly 'natural' parts or 'organs' of the 'normal' body, with the business centre as the head and/or brain, housing and industry as the spine, and heavy industry and factories as the stomach. As Plate 2.2 indicates, the city was zoned into discrete functions and designed to be mass produced and built to high densities. It was an elevated city too with Le Corbusier seeking to develop a continuous park on the surface with building structures raised clear of the ground. In particular, the city, as Frampton argues (1980), was premised on the production of an industrialised norm where compression of space was important in housing the working population. The size of apartments in Le Corbusier's plans for the Radiant City sought to optimise all available space and, as Frampton (1980: 179) notes, 'kitchens and bathrooms were reduced to a minimum'.

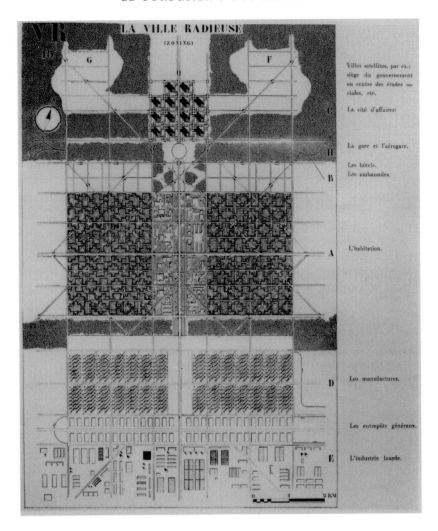

Plate 2.2 Le Corbusier's Radiant City
Source: Foundation Le Corbusier

For disabled people, Le Corbusier's conception of the Radiant City, and architectural forms and processes more generally, was problematical given, as Mumford (1968: 118) suggests, the overvaluation of mechanisation and standardisation as ends in themselves 'without respect to the human purpose to be served'. For Mumford (1968: 119), Le Corbusier 'embraced every feature of the contemporary city except its essential social and civic character'. Others have been similarly critical and Osborn (1971: 420) reacted to the Radiant City by referring to Le Corbusier in the following terms:

37

I hadn't realised till now how fundamentally stupid he was. What is extraordinary is that an architect should have such a lack of the sense of scale and of the feelings of surroundings. He was dominated by a desire for striking visual effects and by a Macaulayan belief in technology and left out everything else.

While such criticisms seem slightly harsh, they identify the essentialist conceptions of Le Corbusier as conceived in specific architectural schemes like the Radiant City. For Le Corbusier, the Radiant City was an abstraction in many ways, but, in particular, the underlying conception of the interrelationship between the body and city was, in Grosz's (1992: 242) terms, 'defacto or external, contingent rather than constitutive'. In this sense, Le Corbusier reaffirmed the primacy of a mechanical conception of bodies.

Reclaiming context and developing self-conscious embodied architecture: concluding thoughts

One of the challenges for architecture and architects is to transcend what Grosz (1995: 127) refers to as 'the standard assumptions, the doxa, the apparent naturalness, or rather, the evolutionary fit assumed to hold between being and buildings'. Such assumptions underpin many contemporary architectural ideas and practices and are rooted, historically, in the theories and conceptions of influential architects and architectural traditions. Architecture has also been beset by major tensions, or, as Crawford (1991: 41, quoted in Imrie 1996: 96) suggests, 'the restricted practices and discourse of the profession have reduced the scope of architecture to two equally unpromising polarities: compromised practice or esoteric philosophies of inaction'. For disabled people, such tensions have worked their way into hostile and oppressive buildings and built environments, underpinned by self-serving ideas which have failed to challenge the hegemonic position of key actors within the wider design and building industries.

As this chapter suggests, architects' conceptions of bodies in space, such as, for example, those propagated by Le Corbusier, are connected to the estrangement of disabled bodies in the built environment. However, it is difficult to derive any easy or definite characterisation of Le Corbusier's feelings and thoughts concerning architecture and the body. In particular, some of his writings hint at conceptions of the human body which are sensitised to the possibilities of flux, change, and difference. In one part of The Radiant City (1967), Le Corbusier conceives of the body as an interconnection between body/mind, that is, the body as a holistic socio-psychological and biological entity:

contingencies are the environment: places, peoples, cultures, topographies, climates . . . contingencies should only be judged as they relate
to the entity – man – and in connection with man, in relation to us,
to ourselves: a biology, a psychology.

(Le Corbusier 1967: 1)

By the end of the 1940s, Le Corbusier (1948: 38) was conceiving of body spaces
in the following terms:

society being in man's image, the nation's wealth in building must be
similar to the human body. Man is carried by a temporal flux in which
he is immersed body and soul; but the flow does more than carry him
physically; it models and remodels him, loosening this, allowing that
to form, operating in a thousand linked ways.

Such views seem to have stemmed from his recognition of the disruptive, contex-
tually specific ways in which clients and/or users of architecture were able to
transform the meanings and materialities of the architects' design conceptions
and intentions. Others have also noted the disruptive interrelationships between
architectural ideas and practices. For Colomina (1994: 126), for instance, 'a
theory of architecture is a theory of order threatened by the very use it permits'.
Moreover, as Tschumi (1996) notes, architecture is never autonomous but is
necessarily constitutive as well as constituted by social processes. For Tschumi
(1996), the fluidity and erratic motions of bodies underpins the possibilities of
new and unexpected spaces being constituted in ways never anticipated by the
architect. Le Corbusier, for instance, recognised the limitations of the application of an architecture premised upon standards determined by logical analysis
and experimentation in noting the subversive nature of people's idiosyncrasies.
As Le Corbusier (1925c, quoted in Banham 1960: 270) argued:

but in practice things do not happen so; man's sensibilities intervene
even in the midst of the most rigorous calculation . . . intervention of
an individual task, sensibility, and passion.

These comments were particularly directed at workers' houses designed by Le
Corbusier in Pessac, south of Paris, in 1925 where he witnessed the inhabitants
transforming the geometric textures and forms of the dwellings (see Plate 2.3).
In interpreting Le Corbusier's contradictory feelings about Pessac, Jencks (1987:
74–5, quoted in Imrie 1996: 86) comments:

Starting with the idea of resolving two incompatibilities like the individual and the group, it was not surprising that Le Corbusier could end up,

as at Pessac, by admiring the way personalisation was destroying his own architecture. All the arguments for a geometrical civilisation . . . were countered by the barbaric actions of the inhabitants at Pessac, and yet, according to the supreme dialectician, these barbarians were still right.

However, Le Corbusier's notion of 'rightness' was still circumscribed by the idea that people, by transforming their living spaces, were somehow, in his terms, subverting and undermining the ideals and purity of the architect. Tschumi (1996: 123), in reinforcing Le Corbusier's observation on the interrelations between buildings and users, notes:

> the human body has always been suspect in architecture: it has always set limits to the most extreme architectural ambitions. The body disturbs the purity of architectural order. It is equivalent to a dangerous prohi-bition . . . architecture, then, is only an organism engaged in constant intercourse with users, whose bodies rush against the carefully estab-lished rules of architectural thought.

Tschumi's views are important for drawing attention to the intersections between bodies and architecture and, in particular, to what he regards as the potentially violent relationship between the two. For disabled people many dimensions of the built environment are disruptive and violent precisely because buildings are underpinned by the embodied ideal of a body which fails to conform with the complexities of bodily interactions in space. Others, in recognising the complexity of such interactions, argue for some reconception of the interrelationship between people and architecture in ways which recognise the fluidity and transformative nature of bodies in space (Grosz 1995; Weisman 1992). For Weisman (1992: 32, quoted in Imrie 1996: 91), for example, universal design offers a direction given its recognition that the built environment should be 'demountable, reason-able, multifunctional, and changeable over time'. Likewise, Davies and Lifchez (1987) conceive of buildings as much more than a physical, bodily, experience, or a matter of logistics, but as a quality of socio-psychological experiences.

This suggests that one pre-requisite for a non-ableist architecture is for architects to confront the social psychology of design by considering the interactions between bodies/minds in the context of specific building use. How do particular buildings and built environments feel to different types of disabled people? Indeed, how do disabled people's feelings interconnect with their bodies' experiences of movement and mobility in specific types of places? What do architects know about this and where do they get their knowledge from and in what form? Moreover, architects need to confront the ideology of art over function and seek to privilege use over aesthetics or pretensions to poetics. This, then, calls, for a demystification and systematic critique

Plate 2.3 Workers' housing at Pessac, 1925 and 1969 showing how Le Corbusier's standardised design was altered over the years by a succession of different occupiers
so subverting, according to Le Corbusier, the original design conceptions
which underpinned the scheme.
Source: Foundation Le Corbusier

of ideas and ideals within particular architectural traditions, theories, and practices
and for architects to (re)connect themselves to social and economic concerns. This
might involve asking who architecture is for, in what ways, and with what effects,
potentially sensitising architects and their clients to the possibilities of architecture
which is inclusive and emancipatory rather than exclusive and oppressive.

The democraticisation of architectural and building practices is also connected to the wider task of developing non-ableist design by moving beyond Le Corbusier's conception of the architect as above and beyond the client and/or user. Few architects in Britain are registered as disabled people while the Royal Institute of British Architects has done little to encourage architects to think about the specific architectural requirements of disabled people (see Imrie 1996; also Knox 1987; Wolfe 1981). However, one also needs to look beyond what Weisman (1992) refers to as failed architecture or prejudiced architects towards the totality of structures framing the social oppression and marginalisation of disabled people within the built environment. Architects are connected to wider cost and material imperatives which inhibit or restrict the scope for design beyond prescribed limits (see Knox 1987). In this sense, an explication of architects' conceptions of bodies in space is only part of a wider endeavour to understand the interrelationships between design theory, practice, and people's experiences of buildings and the built environment.

Acknowledgements

I would like to acknowledge and thank the Economic and Social Research Council which provided grant support (grant number R000235833) to enable me to generate some of the information that this chapter is based upon. My thanks also extend to Anne Boddington for introducing me to a stimulating and rich vein of architectural literature and to Ruth Butler and Hester Parr for some useful insights which were incorporated into the final draft of the chapter. Thanks also to the Fondation Le Corbusier for giving me permission to reproduce Plates 2.2 and 2.3.

Notes

1 The House Builders Federation is a national level organisation that champions the interests of house builders in the United Kingdom. It could be described as a lobby or pressure group that seeks to represent the views of its membership to government and other political parties. It is opposed to facilitating disabled people's access by the construction of specialised and/or additional design features beyond the standard build of housing. It is opposed, so it says, on cost grounds.

2 Where architectural practices consider the disabled body, they tend to by universalising disability as singular and reducible to mobility or ambulant impairments, that is, to the wheelchair user. In this sense, disabled people are seen as a 'type'. Yet, this simultaneously disembodies the disabled person by potentially denying the intersections between the multiplicities of physical and/or mental impairments.

3 As Parr (1997) notes, such ideas hint at the contradictory conception of the body which underlies specific architectural discourses, practices, and forms. That is, the body as seemingly unacknowledged and disembodied and yet particular because it is (conceived to be) of perfect form.

4 The Royal Institute of British Architects (RIBA) is a British-based organisation that confers chartered or professional status on architects who pass through prescribed and validated RIBA architectural training courses. The RIBA also sets codes of conduct for practising architects, sets standards for practice, and provides continuing education and other services to its membership. In teaching trainee architects about disability and architecture, the RIBA's 1993 curriculum briefly mentions disability in Parts 1 and 2 of the Design Studies examination. Students are told that examiners will be looking at the ways in which they have interpreted and worked within the brief which includes taking account of 'disabled movement within the building' (RIBA 1993: 23).

5 Le Corbusier's comments are underpinned by a tension. On the one hand, 'superior mathematical order' might involve 'truths' or a claim to scientific rationality and the predictive capacity of knowledge. However, on the other hand, the 'superior mathematical order' was also a form of pure poetics, beauty, and harmony (a thing of the spirit). Yet, while mathematics and geometric proportion were seen as akin to a form of musical concord for Le Corbusier, and others, they were the underpinnings of 'an entirely objective aesthetic' (Rowe 1975: 51).

References

Ann Hall, M. (1996) *Feminism and Sporting Bodies: Essays on Theory and Practice*, Human Kinetics, Illinois.

Banham, R. (1960) *Theory and Design in the First Machine Age*, Architectural Press, London.

Bordo, S. (1995) *Unbearable Weight: Feminism, Western Culture, and the Body*, University of California Press, Berkeley.

Caygill, H. (1990) *Architectural Postmodernism*, AA, London.

Colomina, B. (1994) *Privacy and Publicity: Modern Architecture as Mass Media*, MIT Press, Cambridge, MA.

Crawford, M. (1991) 'Can architects be socially responsible?', in Ghirardo, D. (ed.) *Out of Site: A Social Criticism of Architecture*, Bay Press, Seattle, 27–45.

Crowe, N. (1995) *Nature and the Idea of a Man Made World*, MIT Press, London.

Curtis, W. (1986) *Le Corbusier: Ideas and Forms*, Phaidon, London.

Davies, C. and Lifchez, R. (1987) 'An open letter to architects', in Lifchez, R. (ed.) *Rethinking Architecture*, University of California Press, Berkeley, 35–50.

Dickens, P. (1980) 'Social science and design theory', *Environment and Planning B: Planning and Design* 6: 105–17.

Frampton, K. (1980) *Modern Architecture: A Critical History*, Thames and Hudson, London.

—— (1991) 'Reflections on the autonomy of architecture: a critique of contemporary production', in Ghirardo, D. (ed.) *Out of Site: A Social Criticism of Architecture*, Bay Press, Seattle.

Ghiraldo, D. (ed.) (1991) *Out of Site: A Social Criticism of Architecture*, Bay Press, Seattle.

Gray, E. (1929) 'De l'eclectisme au doute, l'architecture vivante', translated by Nevins, D., 1981, 'From Eclecticism to Doubt', *Heresies*, 11, 3: 71–2.

Grosz, E. (1992) 'Bodies-Cities', in Colomina, B. (ed.) *Sexuality and Space*, Princeton Architectural Press, New York, 241–54.

—— (1994) *Volatile Bodies: Towards a Corporeal Feminism*, Indiana University Press, Bloomington.

—— (1995) *Space, Time, and Perversion*, Routledge, London.

Hayden, D. (1981) 'What would a non sexist city be like: speculations on housing, urban design, and human work', in Stimpson, C., Dixler, E., Nelson, M. and Yatrakis, K. (eds) *Women and the American City*, University of Chicago Press, Chicago.

House Builders Federation (1995) 'The application of building regulations to help disabled people in new dwellings in England and Wales', unpublished paper, available from R. Imrie, Department of Geography, Royal Holloway, University of London, Egham, Surrey TW20 0EX.

Imrie, R. (1996) *Disability and the City: International Perspectives*, Paul Chapman Publishing, London, and St Martin's Press, New York.

Jencks, C. (1987) *Le Corbusier and the Tragic View of Architecture*, Penguin, London.

King, R. (1996) *Emancipating Space*, Guildford, New York.

Knesl, J. (1984) 'The powers of architecture', *Environment and Planning D: Society and Space* 1, 1: 3–22.

Knox, P. (1987) 'The social production of the built environment – architects, architecture, and the post modern city', *Progress in Human Geography*, 11, 3: 354–78.

Laws, G. (1994a) 'Oppression, knowledge, and the built environment', *Political Geography*, 13, 1: 7–32.

—— (1994b) 'Aging, contested meanings, and the built environment', *Environment and Planning A*, 26, 11: 1787–802.

Le Corbusier (1925a) *The Decorative Art of Today*, Architectural Press, London.

—— (1925b) *La Peinture Moderne*, Paris.

—— (1925c) *Urbanisme*, Cres, Paris.

—— (1927) *Towards a New Architecture*, Butterworth Architecture, Trowbridge.

—— (1947) *The Four Routes*, Denis Dobson Ltd, London.

—— (1948) *The Home of Man*, Architectural Press, London.

—— (1967) *The Radiant City*, Faber and Faber Ltd, London.

—— (1980) *Modular I and II*, Harvard University Press, Cambridge, MA.

Lefevre, H. (1968) *Dialectical Materialism*, Jonathan Cape, London.

Lester, R. (1997) 'The (dis) embodied self in Anorexia Nervosa', *Social Science and Medicine*, 44, 4: 479–89.

McAnulty, R. (1992) 'Body troubles', in Whiteman, J., Kipnis, J. and Burdett, R. (eds) *Strategies in Architectural Thinking*, MIT Press, London, 181–97.

Merleau-Ponty, M. (1962) *The Phenomenology of Perception*, RKP, London.

Mumford, L. (1965) 'The case against modern architecture', in *The History and the City*, New York.

—— (1968) *The Urban Prospect*, Secker and Warburg, London.

Nevins, D. (1981) 'From eclecticism to doubt', *Heresies* 11, 3: 71–2.

Osborn, F. (1971) in Hughes, M. (ed) *The Letters of Lewis Mumford and Frederic J. Osborn: A Transatlantic Dialogue, 1938–1970*, Adams and Dart, Bath.

Parr, H. (1997) Personal Communication, November 20.

Probyn, E. (1993) *Sexing the Self: Gendered Positions in Cultural Studies*, Routledge, London.

Rowe, C. (1975) 'The mathematics of the ideal villa: Palladio and Le Corbusier compared', in Serenyi, P. (ed.) *Le Corbusier in Perspective*, Prentice-Hall Inc, New Jersey, 46–55.

Royal Institute of British Architects (1993) *Examinations in Architecture, Description and Regulations*, RIBA, London.

Sennett, R. (1994) *Flesh and Stone*, Faber and Faber, London.

Shilling, C. (1995) *The Body and Social Theory*, Sage, London.

Sullivan, L. (1947) *Kindergarten Chats and Other Writings*, Wittenborn Shultz, New York.

Tschumi, B. (1996) *Architecture and Disjunction*, MIT Press, Cambridge, MA.

Venturi, R. (1966) *Complexity and Contradiction in Architecture*, Museum of Modern Art, New York.

Vitruvius (1960) *The Ten Books of Architecture*, Dover Publications, New York.

Ward, A. (1993) 'Resistance or reaction? The cultural politics of design', *Architecture and Behaviour*, 9, 1: 39–68.

Weisman, L. (1992) *Discrimination by Design*, University of Illinois Press, Illinois.

Whiteman, J., Kipnis, J. and Burdett, R. (eds) (1992) *Strategies in Architectural Thinking*, MIT, London.

Wigley, M. (1992) 'The translation of architecture: the production of Babel', in Whiteman, J., Kipnis, J. and Burdett, R. (eds) *Strategies in Architectural Thinking*, MIT, London, 240–54.

Wolfe, T. (1981) *From Bauhaus to Our House*, Farar, Straus, Giroux, New York.

3

THE MORAL TOPOGRAPHY OF INTEMPERANCE

Michael L. Dorn

Disability, alcohol and geography 'under the influence'

Disability, like gender, is now recognised as a fundamental, yet long overlooked, category of historical analysis (Baynton 1996, 1997; Mitchell and Snyder 1997; Thomson 1997). Current understandings of disability emerge from a complicated history of religious, legal and medical competition and collaboration over the interpretation and treatment of bodily anomalies. Drawing on scholarship in disability studies and the history of the body, academic geographers have turned their attention to the place of anomalous human bodies in Western society, detailing the production of deviance through practices (scientific, medical, legal, etc.) of exclusion and signification, or rather *stigmatisation* (Dear and Wolch 1987; Gleeson 1996; Parr and Philo 1995, 1996; Philo 1987; Radford 1991). In this chapter, I intend to extend this work by examining the role of socio-medical investigations in giving anomalous bodies meaning. Geographical ideas of environmental medicine and sanitary reform played important roles in the construction of heavy drinking as a social disability in early nineteenth-century United States. During this period, 'intemperate' bodies addicted to alcohol became signifiers of discord in the body politic.

The roots of disability studies in the geography, sociology and human ecology may be traced to the nineteenth-century censuses and social surveys that sought to account for institutionalised and dependent subgroups of the population (Bulmer *et al.* 1991; Driver 1988; Jarvis 1850, 1971; Philo 1995). During this period, high-low hierarchies were deployed by the bourgeois classes as part of a strategy of purification, as the temperate self was defined by its elevated position over and against a seething landscape of dissidence and disgust (Boyer 1978; Stallybrass and White 1986). Popular and scientific understandings of gender, race and human nature were imbricated in these moralising hierarchies; the resultant geographies were written against, but still under the influence of, powerful dissipative forces and flows (McCormack 1998).

Whilst recognising that disabilities are often taken to be objectively present and 'real' by the individuals who experience them, I assert that this sense of reality is always mediated through culturally transmitted understandings of normal bodily appearance and function (Foucault 1972; Thomson 1997). Dominant disability categories are a product of sustained cultural work. The naturalness with which *able* and *disabled*, *temperate* and *intemperate* identities are attached to particular bodies and places depends upon the level of cultural authority commanded by élites, particularly members of the clergy and the professions of law and medicine, and the level of dissidence or mis-fit experienced 'on site'. Representations of disability establish the parameters of the civil and the normal – a form of politico-moral territorialisation. Yet these representations are often also contested by individuals and groups that seek to reclaim behavioural or physiological domains as a method of self-definition, and to normalise behaviour that may previously have been read as deviant. Slight shifts in the frontier between normality and difference indicate broader struggles within and between groups that comprise a given social order.

In some cases, individuals and groups seek to *expand* the coverage of disability categories. Such an expansion took place during the campaign for passage of the 1990 Americans with Disabilities Act (the 'ADA'). Part of the mobilisation effort on the part of disability activists involved forging alliances with groups that had rarely considered themselves 'disabled' previously, including persons living with HIV/AIDS and with mental illness including alcohol and drug addictions. Such an expansion of the category was thought necessary in order to build Congressional support for sweeping civil rights guarantees. Yet the political liabilities of expanding the 'people with disabilities' body count was little discussed at the time (Stone 1981, 1984; Zola 1993).

There are also potential social costs, both at the individuals and group levels, to the over-extension of a medicalised disability status. One case in point would be the rapidly expanding group of children and teenagers being diagnosed with attention deficit and hyperactivity disorders. Public awareness of this 'psychiatric disability' has been actively promoted by pharmaceutical manufacturers who stand to make immense profits due to increased sales of stimulants such as Ritalin. Yet recent critical legal and education research has cast grave doubts on the effectiveness of current approaches to learning disabilities – behavioural patterns that were previously thought to be merely learning *difficulties* – in many American school systems (Metcalf 1998).

Socio-medical observations and classifications of alcohol consumption demonstrate the complicated relations between moral, medical and civil understandings of chronically disruptive behaviour. For example, the ADA prohibits discrimination in public accommodation and employment against persons with disabilities, including persons with alcohol and drug addictions, as long as they are actively pursuing treatment for their conditions. Yet the pool of potential beneficiaries is immense. A recent epidemiologic study of alcohol-related behaviour using the official nomenclature and diagnostic criteria of the American Psychiatric Association found that nearly one quarter of all US men met the qualifications for alcoholism during their lifetimes (Helzer *et al.* 1991). It might not surprise us then, that various states have moved to exclude alcoholics and other substance abusers in their welfare assistance programmes.

The modern disease concept of alcoholism holds that a small subgroup of the population has a (genetic or otherwise implanted) 'allergy' or 'addiction' to alcohol that prevents them from being able to drink in moderation. Developed as a moderate, non-judgemental response to the perceived excesses of national prohibition during the 1920s, the disease concept's focus on 'alcoholism' directed attention away from the physiological effects of alcohol on humans in general (Beauchamp 1980; Roizen 1997). Working with the financial support of the liquor industry, researchers at the prestigious Yale Center for Alcohol Studies set about developing detailed taxonomies of different 'alcoholic' personality types during the 1940s. Once isolated as a medical species, the alcoholics, it was believed, would be freed of the potential stigma of moral failure. Yet medical explanations, codified and popularised in palatable forms like Alcoholics Anonymous' 'Twelve Steps', failed to attain cultural dominance; moral frameworks retained their tenacious grip. Rather than the medical cancelling out the moral, it may be observed that these two discourses merely fed off one another.

The medical and the moral emerged as dual responses to patterns of heavy drinking in early American history. In the colonial period and during the Revolutionary War, the consumption of alcoholic beverages was widespread and accepted. The common opinion amongst the population was that alcohol served an important function in the social order, promoting cohesiveness, and serving both nutritional and medicinal purposes. Whilst members of the Puritan clergy would decry the Devil's influence in those who overindulged, fermented and distilled spirits were still referred to as the 'good Creature of God' (Increase Mather 1673, as cited in Lender 1973). This tolerant approach to liquor would be questioned in the last decades of the eighteenth century. Whilst taverns had served as important loci for the flowering of a protorepublican political culture leading up to the American Revolution, their role afterwards was more ambiguous. Habits of heavy drinking drew intensifying criticism from members of the religious establishment, as well as various cultural commentators, including Benjamin Franklin and Dr Benjamin Rush (Arner 1993; Conroy 1995; Franklin 1987; Rush 1934). The decades following the Revolutionary War saw the emergence of a politicised mass public and widespread concerns about the fate of the American experiment in independence. If the experiment was to be sustained and solidified, it needed to draw upon the ability of patriotic men and women to develop self-control and temperate comportment.

The American alcohol problem was widely recognised by the first decades of the nineteenth century. Missionary reports, letters and biographies of the period are replete with references to men, both young and old, who had been lost to productive labour due to the physiological effects of prolonged alcohol use (Appleby 1997; Drake 1814; Rohrer 1985). Heavy drinking was viewed as the dominant symptom of social disorder. In the wake of a prolonged binge brought on by the resolution of the War of 1812, religious leaders concerned for the physical and moral condition of the young nation inaugurated a campaign to quell the intemperance and moral decline (Rohrer 1985).

At first, their efforts had very little lasting effect. However, by the mid-1820s religious and medical reformers had joined forces. Élite-sponsored societies for the suppression of intemperance that had been common in New England were rapidly replaced by grassroots temperance societies, moral engines patterned on the mutual benefit societies that were so popular amongst the aspiring middle classes (Bernard 1991; Sellers 1991). To legitimate and disseminate their views, these religious leaders relied upon the physiological knowledge and vocabulary of physicians. The activities of physicians were decisive: they catalogued scientific, medical and moral facts concerning the injurious effects of excessive alcohol consumption, and combined them with a detailed social mapping of the urban environment. The product of these efforts was a more comprehensive population health perspective, combined with an activist 'medical police' presence. The mid-century consensus that the consumption of alcohol in any form signified 'intemperance' was tied to the broader triumph of a distinctively American form of able-bodied and temperate male Anglo-Saxonism, shaped by the psycho-social demands of market revolution, democratisation, and intense social mobility.

The role of medical geographic ideas in rationalising this activity deserves further exploration. While environmental historians have examined the work of surveyors and topographical engineers in gridding the landscape and bounding the territory of the American West, they have thus far neglected the ideas of environmental medicine that inspired the activities of physicians and cultural leaders. The leaders of Western society turned to environmental analysis in order to understand the unfolding character of the new Western race, as the original Anglo-Saxon stock faced the challenges of a raw, unfinished environment. Convinced that all disease had environmental causes, early American medical doctors and scientists systematised their knowledge of geophysical and medical domains for the purpose of gaining a more complete view of the population's health. In this chapter, I will examine their efforts to grapple with the increasing social disorder in Ohio Valley cities such as Cincinnati. At first, physicians trained in classical medical geographic ideas deployed humoral analogies in interpreting these social dynamics. Yet in time, the transformed urban milieu required new environmental models. Thus, we find fresh insights from human anatomy deployed in the therapeutic classification and localisation of intemperate and immoral behaviour in Western cities. Out of a previously undifferentiated miasma of human behavioural and physiological imbalances emerged standards of normalcy ('physiological temperance'), as well as categories of human disablement that would become institutionalised by the second half of the nineteenth century.

The moral project of medical topography

Social reformers and social scientists of the early nineteenth century did not maintain the same distinction between physical and moral causes of disease that we do today. Just as the accumulation of filth in the city presented the threat of disease through exposure to 'fetid air', so did the accumulation of criminal or intemperate behaviour create its own 'moral miasma'. The term *moral* is taken to refer to the mores – standards of behaviour or conduct – shared by segments of the population, with an implied hierarchy of value. As Driver (1988: 279) notes, 'the project of social science rested upon the identification of moral regimes and their relationship with environmental conditions. The mapping of the moral geography of the city, in particular, provided a basis for social intervention'. The compilation of the medical topography of a locality – describing the facts of its geology, landforms, drainage, vegetation, climate, customs and industry – was the first step for designing an improved 'medical police' of the region (Frank 1976; Jordanova 1979; Rosen 1974). American doctors writing medical topographies saw no need to differentiate their naturalistic medical topographic observations from their deeply felt personal visions of what American society should become. Professional authority and autonomy were issues of paramount importance, as physicians struggled to displace midwives and folkhealers and to secure the confidence of political and economic élites in the effort to forge an enlightened civil society (Szaraz 1993). Natural history offered doctors the guidance they needed in order to understand the ills of the social body, as they grappled with afflictions such as urban violence and the evils of slavery. Yet while the neo-Hippocratic ideology underlying medical topographic observation proved to be useful in representing the diversity of moral practice, it was still unable to defend the Anglo-Saxon race against eventual degeneration.

The writings of an important early American medical educator show us a great deal about the place of medical geographic discourse in defending Anglo-Saxon racial character and American political union against forces of dissipation. Daniel Drake was one of the best known medical teachers, authors and editors of the United States during the second quarter of the nineteenth century. His best-known contribution to the medical geographic literature is titled *A Systematic Treatise, Historical, Etiological, and Practical, on the Principal Diseases of the Interior Valley of North America, as They Appear in the Caucasian, African, Indian, and Esquimaux*

50

Varieties of its Population (two volumes, the first published in 1850, and the second posthumously in 1854). This immense, over 1400 page work, based on Drake's own observations made during summer travels during the 1830s and 1840s, and supplemented where necessary by facts supplied to him by a large circle of correspondents, attempts to present a picture of all of the characteristics of the inhabited portions of the American West that can help to explain the patterns of health and disease as experienced by its population.

The first volume of Drake's *Treatise* (1850b) presents the facts relative to the topographic, hydrographic, climatic, physiologic and social environmental that may help account for perceived patterns of disease emergence. The third section on physiological and social etiology covers the migration of different racial and ethnic groups into the Valley, the general characteristics and variation of the diet (food, thirst-quenching beverages and stimulants) and the various habits (clothing, lodging, bathing practices, occupations, recreations, etc.) that relate to health outcomes. Each of the world's races, associated with one of the four winds, bring their phlegmatic or lymphatic ('Esquimaux'), melancholic ('Caucasian'), sanguineous ('Indian'), and bilious or choleric ('African') temperaments to the great mixing bowl (for typology, see Haller 1981: 5). Drake lays this humoral schema over a hydrographic map of North America, describing the human tide descending in waves into the Valley of the Mississippi. The characteristics of the new commingled race – the 'Westerners' with whom Drake always identifies – are still being formed. The dominant Anglo-Saxon type is to be refined and improved in this, the '*last* crucible into which living materials, in great and diversified streams, can be poured for amalgamation' (Drake 1850b, reprinted in Shapiro and Miller 1970: 349). The new habits being exhibited by residents of the Interior Valley are described in the most forthright fashion, so that physicians referring to Drake's treatise will know where to intervene for the benefit of this and future generations. Singled out for denigration is the widespread use of alcoholic beverages and tobacco. Drake believed that only by careful observation while it was still young and ductile could the 'Western' mass be trained into upright, healthy and temperate citizens.

The physician in the secular American republic assumed an important responsibility for organising and diffusing knowledge that the population needed to avoid moral and physical degeneration. During research trips gathering data for his *Systematic Treatise*, Drake frequently related his latest discoveries in letters to his medical colleagues. These letters found their way into the hands of the editors of Cincinnati and Louisville medical journals. Often the letters included moral tales, discursively defending norms of civil and healthful Anglo-Saxon behaviour against the influence of grotesque, degenerate forms, associated with the 'lesser' races.

In one such letter, written while on an extended journey through the mountains of Western Virginia, Drake (1850a: 568) describes the habits of a tribe of

avages known as the 'Hinkleites'. Upon entering a remote mountain
ke happens upon the encampment of a 'peculiar tribe of vagabonds'.
nty or thirty in number, they stand, sit, or even lie on ground that is
om a heavy rain the previous evening. Drake presumes that he is observ-
l of emigrants or gypsies, before learning from 'intelligent persons' at
the next village of the group's Anglo-Saxon background. 'They are content with
the crudest diet and the vilest whiskey, provided it will but make them drunk . . .
and have no confidence, when ill, in the medical profession; but "Doctor" themselves with "roots and herbs"' (Drake 1850a: 569). Drake ascribes to this Virginia
family the same popular enthusiasms that bedeviled his professional life back home
in the cities of the Ohio Valley: the general intemperate consumption of food and
drink; and trust in quack medicines and sectarian healers.

The wild Hinkle children of the mountains, insufficiently differentiated from
their ecosystemal context and capable of withstanding extremes of rain and cold
with little protection, afford 'a striking example of the adaptation of the physiology of the white man to that which seems the natural inheritance of the red'
(Drake 1850a: 569).[1] For Drake, environmental analysis holds the key to understanding health and disease. The medical geographer might prescribe a change
in environment as the first step towards health. Drake's environmentalism allows
for the possibility that persons of Anglo-Saxon heritage exposed to the North
American climate on a daily basis would become like the Native Americans.
Their environment fails to provide them with the sort of social, cultural and
climatological 'seasoning' necessary to develop physiological temperance (Drake
1842a). As an Ohio Valley physician, Drake set out to document and counteract
dissipative forces that endangered the Western racial stock. Yet in order to
accomplish his goals, Drake would be forced to abandon his topographical
approach for a more explicitly anatomical one. In place of the humoral economy
of flows that stressed moderation and balance, Drake would come to advocate
a more rigid and retentive economy of abstinence. The body as a citadel of
health was to be defended from legions of barbarous outsiders, marked by their
licentiousness and lack of self-control.

Marketisation, democratisation and the turn to abstinence

The Hippocratic writings on dietetics constitute a course in techniques of the
self (Foucault 1985). Proper control of the appetite for moral and physical stimulants requires a 'regimen', adapted to both the balance of internal energies and
the influences of the physical environment. Knowledge and dissemination of
these techniques are an important part of the medical profession's approach to
disease, for one cannot treat individuals without rectifying the lifestyle that makes

them sick in the first place. The topics covered in Book IV of the Hippocratic work *Epidemics* include exercises, foods, drinks, sleep and sexual relations (Hippocrates 1983: 6). In order to understand the emergence of a new regime of normalcy (and the creation, by process of elimination, of numerous 'abnormalities') in the early nineteenth century, I will be examining the changing medical regimen of 'stimulation' as it shifts away from the temperate self-fashioning advocated in Greek dietetic literature. Medical ideology during the 1820s and 1830s, while still based on classical ideas, was realigned towards the specific problematisation of intemperance *vis-à-vis* alcohol. Eventually, a new set of social forces led medical doctors and religious leaders to sacrifice Hippocratic moderation and balance for a proto-Victorian regime of self-denial and abstinence.

Throughout his young adulthood, Drake demonstrated little interest in the repression of the passions. From his earliest writings on the beneficial effects of voluntary societies on their members, he maintained that youthful energy and ambition are to be expected in young men (Drake 1807). Rather than being expended in dissipation on luxurious and passionate living, these should be diverted into more useful channels. All humans possess a taste for, and propensity to take advantage of, stimulation. Problems only begin when this taste for stimulants is misdirected:

> As long as man stimulates himself moderately, and with such articles as do not impair his health, or pervert the faculties of his soul, he violates no moral or physical law, and suffers no immediate or prospective injury; but the moment he selects and indulges in such as do either, he is a transgressor, and must suffer the penalty of his violation . . . Philosophy and ethics do not, then forbid all stimulation; but occupy themselves in regulating the selection and quantity of stimulants.
>
> (Drake 1831, reprinted in Shapiro and Miller 1970: 206)

The regime is a mode of problematisation that endeavours to save the desire for stimulation from its inappropriate satisfaction (Foucault 1985: 101). For Drake a number of different stimulants must be regulated – foods and beverages, exercise and amusements – but only certain of these become sufficiently problematised that a campaign is organised to reform practice. Amongst the salutary stimulants are tea and coffee, cider, beer and the milder wines. Drake distinguishes the consumption of these 'natural' stimulants from the consumption of ardent spirits, which were a product of 'artificial' distillation. The locations of endemic artificial stimulant consumption are the proper objects of Temperance scrutiny:

> The causes which give activity to the propensity for stimulants, are many and diversified. Some are moral, others physical. A part are

universal, but the greater number are local and special; operating in particular places, or on certain groups of society. It is to these *causes,* that the friends of Temperance should direct their attention. Prevention should be the object: drunkenness is seldom cured, but has often been averted; and will continue to disappear, in the ratio in which its causes are laid open and rooted out.

(Drake 1831, reprinted in Shapiro and Miller 1970: 208)

During the 1820s and 1830s, Drake's hometown of Cincinnati, Ohio, experienced a precipitous rise in these local causes of drunkenness. The increased problematisation of alcohol consumption became both possible and useful in Cincinnati at this time for two reasons: the diffusion of voluntary association, and the fruition of market revolution. Voluntary associations arose as a replacement for the unravelling social order of an agrarian-based community that had traditionally been responsible for moulding young people (men in particular) for active and moral citizenship. These new societies for mutual benefit exerted their own homogenising influence.

Among the emerging middle classes of the nineteenth century, from whom traditional group affiliations had been shorn, the desire for identity produced conformity that was expressed in an intolerance of differences – precisely those distinctions that freedom encouraged. Because democracy precluded former class alliances and generational continuities, people had only one another after which to model themselves.

(Thomson 1997: 43)

These new models originated as a response to frustrated social ambitions.

The gap between social ambitions and the means to express them grew with the arrival of the market revolution, carried into the Midwest on the heals of improvements in canal and road travel. After the opening of the Miami Canal on 17 March 1828, the residents of Cincinnati were soon swept up on a surge of optimism as the new transportation network connected the entire Ohio River to the Great Lakes in the North and eventually, via the Erie Canal, to eastern markets. Commercial farming in the Miami Valley region boomed and increased amounts of corn, pigs, flour, wheat and whiskey moved through Cincinnati. The prosperity of the region also brought a new tide of immigration, particularly from Germany. The city's population grew from 16,000 in 1826 to 46,000 in 1840, making it the fastest growing major city in the West. Thirty per cent of that total was German (Starr 1982: 47; Ross 1985).

This is also the period when Cincinnati earned a reputation for liquor and pork production. Some local wits even suggested that its nickname should be changed from 'the Queen City' to 'Porkopolis'. Farmers growing corn in the fertile soil of the Ohio Valley saved themselves the difficulty of carrying bulk grain to market through Cincinnati to New Orleans, where it might rot on the wharves or sell for less than the cost of the trip, by either feeding it to pigs or distilling it themselves into whiskey. In the process, a corn glut was converted into a pig and whiskey glut (Rorabaugh 1979: 76–81; Dannenbaum 1984). While both pigs and whiskey generated similar profits, whiskey's advantages included its ease of storage and increased value with age. Cincinnati became the regional focus of the drink trade, offering whiskey ware-housing, wholesaling and shipping. An urban topography of spirits production and consumption emerged by 1834, as merchants rich on the whiskey trade built opulent houses in the hills and 223 saloons and taverns adorned the city, or approximately one for every thirty-eight men aged 15 and over. The municipal government was also get-ting fat on the annual income from liquor licence renewal: $9,682 or 18 per cent of the city budget in 1834 (Dannenbaum 1984: 24–5).

By the beginning of the 1830s, not just Cincinnati but the entire nation seemed to be on an extended drinking spree. The historian William J. Rorabaugh (1979) refers to this period as 'the alcoholic republic' due the unparalleled levels of distilled spirit consumption that accompanied the arrival of Andrew Jackson to the American presidency. Newspapers, trial reports and popular autobiographies of the period recount sensational stories of Western men, addicted to liquor, abusing or murder-ing their families (Nadelhaft 1987; Cohen 1995). Adverse reactions to the alcohol withdrawal, known as *delirium tremens*, *mania à potu*, or *mania à temulentia*, became a commonplace medical diagnosis, and drink-induced insanity was argued as a defence in the courts (Hayward 1822; Drake 1830; Ray 1838; Rorabaugh 1979). While the erratic behaviour of men was of principle concern, women were also known to reg-ularly overindulge.[2] Members of the 'fairer sex' were thought to be more quickly and certainly ruined by the practice (Drake 1828).[3]

In 1830 the average American over the age of 14 drank 9.5 gallons in one year. This period ranks amongst the highest rates of alcohol consumption for any country at that time, and approximately three times the present level of US consumption at the close of the twentieth century. Yet by 1845 consumption of hard liquor was almost a third of its 1830 level, while the amount of absolute alcohol consumed had fallen to nearly a quarter. What had happened to bring about this sudden reversal?

Responding to the widespread belief that the American experiment was being squandered for a collective debauch on ardent spirits, religious and medical reformers joined forces to redesign élite-sponsored societies for the suppression of intemperance into moral engines to sustain the temperate (Bernard 1991; Sellers 1991). In order to diffuse temperance discourse throughout all the new voluntary associations being created for self-improvement, the message had to be both personal and simple. Picking up on the self-help tenor of artisan organising, Lyman

Beecher's *Six Sermons on Intemperance* helped shift the discourse of the American Temperance Society away from the suppression of intemperance (as immoderation), towards the promotion of total abstinence for all (Sellers 1991: 263).

This radical aim was initially considered shocking to the conservative establishment of a country where alcohol consumption was both traditional and pervasive. Such an appeal was bound to displease the élites who financially backed the first temperance societies in the North-east, and who maintained that their own preference for wine was not related to the social problems of dissipated masses. Such an appeal would also alienate the Church leaders who resented the aspersions cast on their sacramental wine. Instead, the message of abstinence was calculated to engage the spiritual and financial destitution experienced by so many men and women disoriented by the unfettered marketisation of American life. For the small businessmen and artisans who belonged to mutual benefit societies or sought bank credit for new ventures, abstinence was a mark of thriftiness and high character (Dannenbaum 1984: 22). For others looking for reprieve from personal hardship, abstinence served the same social function as the religious fast, both as a form of penance and as a mark of personal commitment to a new, quasi-religious group identity (Bernard 1991).

Casting new light on the alcohol trade

With the arrival of the market revolution, labourers, artisans and small businessmen in the Ohio Valley were increasingly vulnerable to the cyclical booms and busts of the capitalist business cycle. While most got by during the economic boom of the mid-1930s, the nationwide economic depression that began in 1837 was devastating. It was a constant struggle simply to make ends meet: weekly wages in Cincinnati were cut, and, as wives, daughters and younger sons went to work to make up the difference, the worst fears of members of the producing classes – poverty and dependency – were realised. In this new economic climate it became very difficult for labour activists to identify specific sources of distress that might be remedied. As a result, worker street demonstrations and organising gave way to individualist remedies – self-denial and appetite repression carried out on a mass scale (Ross 1985).

The years of hardship that followed the Panic of 1837 set the stage for a new blaze of temperance light. Millions of physically and morally defeated men and women looked to a grassroots movement of reformed drunkards named after the first American President. The Washingtonian Temperance Movement (a forerunner of today's Alcoholics Anonymous) coursed through the countryside of the 1840s with millennial intensity. The spread of Washingtonian light exposed a society undergoing far-reaching changes. Visual evidence of an urban topography arrayed with social ills cast doubt on the sanctity of representative democracy, based on nature's laws and protected by a Divine hand from the degeneration found in the European working classes. Imagery of Cincinnati as a kind of Republican utopia of industry and temperance was no longer believable (Drake 1815). While maintaining his lifelong contempt for 'High Churchism', Drake took communion in the Protestant Episcopal Church during the early months of 1840. Evangelical Christianity provided a new vehicle for his moral energies, as he organised weekly prayer meetings and temperance society meetings for medical students. His newfound asceticism also accorded with the ideology of abstinence promoted by the Washingtonians.

The first Washington Temperance Society was founded on 4 April 1840 by six drinking buddies at a tavern in Baltimore who decided, after discussing a temperance lecture that four had attended, that they should form their own mutual benefit association (Blumberg and Pittman 1991). The next day it was decided that they should each quit drinking and together start a society for total abstinence. As a society created by and for reformed drunkards, they decided that they would devote a portion of each meeting to testimonials by members describing their former state and the benefits that have accrued from their newfound temperance. Meetings organised for public audiences that November proved to be very popular, and by their first anniversary there were over 1,000 drunkards and 5,000 other members and friends to celebrate in a parade down the Baltimore streets. By the spring of 1841, delegations were being sent out to spread news of the reform to other cities. Before long, the wildfire of Washingtonian temperance was blazing across the Mississippi Valley; the first meeting of the Cincinnati Washingtonian Society took place that July. Whilst out West, travelling in pursuit of data for his *Systematic Treatise*, Daniel Drake (1842b: 394) noted the progress of this new light across the countryside:

> We have lately made personal observations of the Temperance Reform from St. Louis, Missouri to Chillicothe, Ohio, and are enabled to state that an extraordinary reduction in the use of intoxicating drinks has taken place within the last six months. Temperate drinkers have ceased to indulge themselves, and a vast number of habitual drunkards have suddenly broken off, and joined in the crusade against the use of alcohol.

Washingtonian societies (and their female-led Martha Washingtonian counterparts) challenged generally held assumptions about alcohol addiction.[4] Generations of reformers had maintained that their responsibility was to prevent the erosion of good habits in the temperate. Once a pattern of intemperance was established in the drunkard, it could not be modified – it was a road to almost certain death. Yet here was a movement formed *by* reformed drunkard *for* fellow drunkards. This development gave Drake and others new confidence that the front lines of reform might be pushed beyond the territory occupied by the voluntary associations.

Rather than leading by conspicuous example, temperance activists carried their message to recruiting sites for the opposing army. Western cities were becoming fertile breeding grounds for the forces of dissipation and disorder rather than centres for emulation and improvement, where individuals of different backgrounds joined together in cooperative effort to meet their communal obligations. As new immigrant groups brought their foreign ways to the city, they were no longer uniformly bent towards the needs of society as a whole through the effects of crucibles like the schoolroom or the church. Instead they fell under the sway of the forces of disorder. According to the unpublished Drake address 'Breweries vs. Foundries' (1841), the army of the intemperate was growing in strength, drawing fresh recruits from all strata of society, from the stately homes of the wealthy and the hovels of the destitute:

And who are the recruiting officers of this motley multitude? I will tell you. They are of every grade, from him who may be regarded as the Commander on the recruiting service, down, through sergeants and corporals, to the drinking profligate and vagabond-recruit, who is sent out as a decoy.

1. **The fine gentleman**, who displays the advantages of wealth, education and refinement, in spending his afternoons over the wine bottle . . . while he declaims against the use and abuse of whiskey and porter by the poor; and denounces the bigotry of Temperance Societies.

2. **The elegant and fashionable lady**, who at the hour of midnight assembles her young friends – of both sexes – round her ostentatious and luxurious table, to reanimate their *drooping* spirits, with *ardent* spirits, in the guise of champaign!

3. **Young men of fortune**, who take one glass with their fathers, and complete the days drinking in the saloons or a coffee house, winding up at 11 o'clock paying the bills of those who have drunk themselves poor and who they thus draw in as companions and flatterers.

4. **Young ladies with tastes which enable them to enjoy the society of young gentlemen**, whose breath sends up the mingled stench of Tom & Jerry, whiskey punch and squealing pig!

5. **The Demagogue**, who holds the corruption of the people but a minor matter compared with his election to office.

6. **The merchant** who keeps his wine and brandy locked up in his counting room, for his own temperate use, and tells his clerks that they must not drink till they get into business . . . Admirable consistency.

7. **Master workmen**, especially of those trades, which assemble apprentices, journeymen and operatives in the same shops; where they can talk and laugh and drink . . .

8. **Town and City Corporations**, who by becoming drinking houses declare that they are of public utility; and promote drinking by multiplying temptations. These establishments, at once the cause and effect of intemperance, should be taxed only to be suppressed.

9. **The steam boat owners and commanders** who keep bars and drinking and gambling saloons on their boats . . .

10. **Indian traders** who clandestinely violate the beneficent laws of the land; and for the sake of gain counteract the government and benevolent and kind hearted people who are labouring to bring the red man from savage degradation.

11. **The distiller and adulterator** who make *tasters* of their sons and operatives . . .

12. **The brewer of Ale, porter and beer,** whose bottles and kegs are hawked around our towns and cities in a cart . . .

13. **Tavern keepers**, who would throw up their licenses if drams at their counters could no longer be sold; and who think much more of accommodating the tipplers of the city, than orderly and temperate travellers, who desire lodgings with them.

14. **Managers of the Theatres**, who promote or permit the establishment, around and within their edifices, of drinking houses, in which the profligate of both sexes congregate . . .

15. **Keepers of gambling houses** . . .

16. **Keepers of coffee houses**, where the poisonous bowl is wreathed with roses . . .

17. Lastly, **keepers of Doggeries**, where loafers are kennelled at night on a naked and filthy floor, to which the coffee house customer begins to resort, when the means of more genteel indulgence are wasted and gone – where he can get drunk at half the price – and whence he is sooner or later taken away by the Observer of the Poor, in a two dollar pine coffin stained of the same colour with the carbuncled face, to be buried at the public expense.

Source: Drake, D. (1841) 'Breweries vs. Foundries; Address given at Louisville, Kentucky, December 25, 1841 and Chillicothe, Ohio, July 4, 1842', from Daniel Drake Collection, Mss qD761 RM, Cincinnati Historical Society, Cincinnati Museum Centre.

An educator of medical students and high profile advocate for the temperance cause, Drake became a moral role model for many men of the Ohio Valley. It was Drake the father and grandfather that was so anguished to observe a growing class of 'bullies and desperados', and the 'ring-streaked, speckled, and spotted' – the gross and sensual 'votaries of dissipation' (Drake 1847). To protect the health and the souls of these men, he spiced his lectures with dietetic and moral regulations, and offered additional moral instruction to the members of the 'Physiological Temperance Society' (Drake 1842b). He consistently advocated that young men devote themselves to temperance. As an authority on moral regulation, he would also receive letters from men who, in the Washingtonian mould, detailed the story of their fall into inebriety and the benefits that they derived from current sobriety (Drake 1851).

A generalised state of anxiety was brought on by the destructive effects of the market revolution on agrarian social values: 'Torn loose from the moorings of traditional society with no clue as to how to behave, thousands of people were literally thrashing around in the early republic. The mixture of high ambitions, limited resources, and cultural confusion made frustration and disappointment common experiences' (Clayton 1986: 47). As competitive merchant capitalist social relations rapidly become the norm in the urban market-place, more genteel and refined modes of behaviour had taken refuge in semi-public, liminal spaces: the library, the drama theatre, the dance parlour, the bookstore, the university. This network of refuges, or 'asylums', formed a kind of buffer for the upper class from the emerging dissipation and rowdiness of the saloon, the tavern and the riverfront market. This social distance allowed a form of status politics to appear in urban centres like Cincinnati, Ohio.

For the asylum-seekers, intemperance constituted a moral plague on the Anglo-Saxon race, potentially just as lethal as cholera. While the recruiters and recruiting sites from which this miasma emanated would modify over time, it would not be alleviated by the standard treatments. Only by eliminating the alcohol trade would one be eliminating the most fruitful cause of disease. Drake saw in the Washingtonian movement a curative force that only needed to be directed to the proper sites. Just as improvements in sanitation would open up pathogenic urban sites to a dissipating hygienic light, so would improvements in temperance rhetoric shed light on sites of moral degradation.[5] Later generations of public health officers, more statistically inclined than Drake, would continue to root out these deformed sites in the body politic.

Conclusion: from the natural to the normal body

Beginning in the mid-nineteenth century, the way physicians looked at the objectives of treatment began to change. They came to think of disorders less as systematic imbalances in the body's natural harmony, and more as complexes of discrete signs and symptoms that could be analysed, separated and measured in isolation. The Paris school's emphasis on the anatomical localisation of diseases introduced some measure of such analysis, but the conceptual shift in the third quarter of the century went much further. The hallmark of the new way of thinking about the goals of treatment was the reduction of signs of bodily order and disorder to objectively measured, quantified norms.

(Warner 1986: 86)

Numerous historians of life sciences and of disability have pinpointed the first half of the nineteenth century as a period of epistemic rupture, the emergence of new discourses of embodiment, from an order of the 'natural' to an order of the 'normal' (Warner 1986; Canguilhem 1989; Pickstone 1994; Baynton 1996). With the rise of clinical observation and quantification, the physiological differences previously ascribed to imbalances in the four humours now figured as *deviations* from a posited mean. For the French mathematician Adolphe Quételet, moral facts demonstrated the same regularity as physical ones. 'Man, without knowing it, was subject to divine laws' revealed through the workings of probability (Rabinow 1989: 66; Davis 1995). Through the advances in pathological anatomy introduced by the Parisian philosopher–surgeon Xavier Bichat, life was now understood as the sum of all forces that oppose death; the 'normal' only emerged through examination of physical differences for the 'abnormal' and 'pathological'.

Similar shifts can be detected at the level of the body politic: the image of city life that emerged in the Ohio Valley after 1840 was less topographical than anatomical: a natural system, or organism, with its own circulation and organs (Figlio 1976; Foucault 1980; Davidson 1983). From this perspective, 'disease is no longer a bundle of characteristics disseminated here and there over the surface of the body . . . it is a set of forms and deformations, figures and accidents and of displaced, destroyed, or modified elements bound together in sequence according to a geography that can be followed step by step' (Foucault 1974: 136). By drawing up a taxonomy and hierarchy of recruiters for the Brewer, Daniel Drake identifies the outlines of an entire occupying military force out of what at first appeared to be simply an unruly mass. This new way of viewing the city, what Mary Poovey (1995) calls 'anatomical realism', is indebted to an emerging transformation in the way Americans viewed the social realm.[6]

The revolution of clinical observation creates a whole new vocabulary, now available as metaphors to be applied towards developing a better understanding of the topographical realm as well. 'The city with its spatial variables appears as a medicalisable object. Whereas the medical topographies of regions analyse the climatic and geological conditions which are outside human control, and can only recommend measures of correction and compensation, the urban topographies outline, in negative at least, the general principles of a concerted urban policy' (Foucault 1980: 175). The body politic is now available as a site for systematic investigation and patient documentation. As the high priests of health, physicians help articulate this new vision of the city with an interiority that must be explored and catalogued. Over the decade 1840–50, the submerged continent of vice begins to emerge and be mapped into different territories under the watchful eyes of new police organisations. New powers of government are mobilised by this vision of a city under permanent attack by a plague of 'unAmerican' behaviour (Marcus 1991).

By examining the genealogy of modern medical geography, one arrives at a new understanding of the place of disability in geographical discourse. In the writings of Daniel Drake we see detailed attention being paid to both the physical and moral components of health, from a broadly ecological perspective. His early writings demonstrate a healthy respect for the variety of health practice in the West, and his initial involvement in temperance activism is a natural outgrowth of these responsibilities. Yet over the decade 1840–50, Drake's empirical geographic vision begins to articulate much more closely with the needs of government and the regular medical profession. His moral topographies of intemperance exhibit a new intolerance for alternative modes of living and behaving in the city. In accordance with this new mode of representation, new 'deviant' sub-groups are soon carved from the mass of urban poor and constituted as objects of surveillance: the drunkards, the deaf and blind, the cripples, the lunatics and idiots.

While geographical practices assist in the ordering and management of this new landscape of dissidence, over time its categories become naturalised, their constructed nature forgotten. The apparent absence of disability themes in modern geographical discourse is then no mere oversight. Typically the process of identity formation for geographers depends upon the rigid exclusion and anatomisation of 'abnormal' viewpoints. Geographical expertise, spatial cognition and other terms of professional qualification are valorised through the denigration of a constitutive outside of errant tropes: the intemperate delusion, the queer frolic, zombie-like trance, the nomadic ramble, the disorderly mental map that is 'contorted, folded, torn' (Vujakovic and Matthews 1994; see also Dorn 1998). Until recently, geographer's excursions into the domain of disability have typically displayed an astonishing callousness to the politics of naming (Dorn 1994; Parr and Philo 1995). This should not be surprising, since adopting diagnostic criteria and imagery from the disabling professions is one means by which academics establish their own professional credibility (Illich *et al.* 1977; Foucault 1979; Pile 1996). In order to advance an emancipatory project, it is important that geographers acknowledge the spatio-temporal structuring of the definition(s) of disability that they work with, and that possibility that there are alternative ways of structuring this experience as well.[7]

Acknowledgements

Thanks go to RGS-IBG G.H.R.G. for a travel grant to attend the medical geographers conference in Portsmouth, UK (1996), where this essay was originally presented. Thanks also to R. Butler, V. Del. Cascino Jr, H. Parr and J. Pickles.

Notes

1 For an extended study of how high/low dichotomies were used in articulating moral topographies in eighteenth- and nineteenth-century England, see the work of Peter Stallybrass and Allon White (1986).

2 All available evidence suggests that levels of alcohol consumption amongst Western males far exceeded that of females during this period (Rorabaugh 1979; Gomberg 1982). Men typically drank more often during the day, and in larger quantities. Temperance lecturers, invariably male, focused on the tragic inability of young men to provide for their families. The wife of the drunkard carried the heaviest responsibility in the temperance reform; she would exercise moral persuasion in the domestic sphere (Dannenbaum 1981). 'Should her husband's vices obtrude upon her narrow and rugged way, she must effectually dispel them, or relinquish hope save that of Heaven' (Drake 1828: 55). Nevertheless, intemperate drinking by women was becoming a fruitful source of concern in its own right, particularly when it contributed to the loss of respectability and the abuse or abandonment of children (Alexander 1988). According to the ladies who ran the Cincinnati Orphan Asylum, many of their charges were the off-spring of respectable parents, who had been brought down by an unquenchable thirst for ardent spirit (Black 1952: 109).

3 To make this point, Drake (1828) translated from French and appended to his *Discourse on Intemperance* a fantastic set of European medical case reports, detailing in gruesome detail the apparent 'spontaneous combustion' of older women with confirmed habits of intemperance. This publication was acknowledged by Drake's colleagues; one even reported a similar case from his Kentucky practice (La Roche 1830; Short 1830).

4 Women had been active in the temperance movement from the mid-1820s establishment of the American Temperance Society. By 1831, one national census found twenty-four upstart 'ladies' temperance societies, while, according to Ian Tyrrell (1982: 131–2), women composed 35 to 60 per cent of the members of male-led societies during the 1830s. The Washingtonian Reform of the 1840s drew larger numbers of Methodist, Baptist and evangelical Presbyterian women, many from lower-middle and even working-class backgrounds, and afforded a few unprecedented opportunities for developing organisational and public speaking skills (Johnson 1843; Tyrrell 1982; Alexander 1988).

5 Responding to the growing mass of dissipated poor in the urban environment in Cincinnati, Drake campaigned for the removal of one particular source of irritation.

Since the opening of the Commercial Hospital in 1820, a poorhouse had been located in the basement of the facility. This had become a source of frustration for the administration, as the inmates downstairs would frequently disrupt the operation of the hospital itself. In 1845 Drake led a citizen's movement to solicit subscriptions and convince the city council to build a 'house of reform and education' in the countryside north of the city. Over time, the plans expanded to include both the poorhouse farm where the able-bodied poor were to learn habits of industry, and the house of refuge, a school of reform for wayward youth. These facilities, completed in 1850, were to provide asylum for the intemperate and dissolute from the pernicious influence of their old haunts and associates, and expose them to the salubrious effects of an exurban, agrarian environment (Marcus 1991).

6 Alan I. Marcus (Marcus 1991: 39–40) cites a diverse group of historians in support of his contention that a shift in ideas about the nature of American society occurred around 1840, with distinct institutional ramifications. Even though Marcus does not cite them, it should not be difficult to find work by historical geographers that also supports this contention. As one important example, Richard A. Walker (1981) develops his theory of suburbanisation around a Marxist periodisation of cycles of capitalist accumulation. These periods approximate fifty-year Kondratieff cycles and are associated with distinctive urban geographies. 1840 demarcates the transition from the mercantile space-economy (c. 1780–c.1840) and the industrial space-economy (c.1840–c.1890). While this transition takes on distinctive characteristics in different localities, the broad outlines of Walker's model hold roughly true. Few other aspects of American society had such a pervasive influence on individual well-being as boom and bust cycles of the emerging capitalist economy, bringing about profound changes in behaviour and discourse (Sellers 1991). It appears that new ideas in science and government articulated with these new economic conditions to change the discursive field (Dorn 1994). 'That a large segment of the American populus seems to have held similar notions of social organisation at the same time implies that those notions may well have been at work in a wide variety of human endeavors, not just city governments' (Marcus 1991: 39).

7 This was one of the positive lessons that disability geographers can gain from the often quite heated debates in *Transactions of the Institute of British Geographers*. The assumptions underlying the behavioural geography approach advocated by Golledge (1993) – research *for* the disabled – were attacked by a younger generation of geographers advocating research *with* and *by* the disabled (Butler 1994; Gleeson 1996; Imrie 1996). As Golledge (1997: 391) notes, without a basic recognition of differences in approach concerning the nature of disability, our research runs the risk of seeming futile or irrelevant.

References

Alexander, R. M. (1988) '"We are engaged as a band of sisters": class and domesticity in the Washingtonian Temperance Movement, 1840–1850', *The Journal of American History* 75, 3: 763–85.

Appleby, J. (1997) 'The personal roots of the first American temperance movement', *Proceedings of the American Philosophical Society* 141, 2: 141–59.

Arner, R. D. (1993) 'Politics and temperance in Boston and Philadelphia: Benjamin Franklin's journalistic writings on drinking and drunkenness', in J. A. L. Lemay (ed.) *Reappraising Benjamin Franklin: A Bicentennial Perspective*, Newark, DE: University of Delaware Press.

Baynton, D. C. (1996) *Forbidden Signs: American Culture and the Campaign Against Sign Language*, Chicago: The University of Chicago Press.

—— (1997) 'Disability: a useful category of historical analysis', *Disability Studies Quarterly* 17, 2: 81–7.

Beauchamp, D. E. (1980) *Beyond Alcoholism: Alcohol and Public Health Policy*, Philadelphia: Temple University Press.

Bernard, J. (1991) 'From fasting to abstinence: the origins of the American Temperance Movement', in S. Barrows and R. Room (eds) *Drinking: Behavior and Belief in Modern History*, Berkeley: University of California Press.

Black, R. L. (1952) *The Cincinnati Orphan Asylum*, Cincinnati: Robert L. Black.

Blumberg, L. U. and Pittman, W. L. (1991) *Beware the First Drink: The Washingtonian Temperance Movement and Alcoholics Anonymous*, Seattle, WA: Glen Abbey Books.

Boyer, P. (1978) *Urban Masses and Moral Order in America, 1820–1920*, Cambridge, MA: Harvard University Press.

Bulmer, M., Bales, K. and Sklar, K. K. (eds) (1991) *The Social Survey in Historical Perspective 1880–1940*, Cambridge: Cambridge University Press.

Butler, R. (1994) 'Geography and vision-impaired and blind populations', *Transactions of the Institute of British Geographers* 19: 366–8.

Canguilhem, G. (1989) *The Normal and the Pathological*, New York: Zone Books.

Clayton, A. R. L. (1986) 'The failure of Michael Baldwin: a case study in the origins of middle class culture on the Trans-Appalachian Frontier', *Ohio History* 98: 34–48.

Cohen, D. A. (1995) 'Homocidal compulsion and the conditions of freedom: the social and psychological origins of familicide in America's early republic', *Journal of Social History* 28: 725–64.

Conroy, D. W. (1995) *In Public Houses: Drink and the Revolution of Authority in Colonial Massachusetts*, Chapel Hill, NC: University of North Carolina Press.

Dannenbaum, J. (1981) 'The origins of temperance activism and militancy among American women', *Journal of Social History* 15, 2: 235–52.

—— (1984) *Drink and Disorder: Temperance Reform in Cincinnati From the Washingtonian Revival to the WCTU*, Urbana, IL: University of Illinois Press.

Davidson, G. (1983) 'The city as a natural system: theories of urban society in early nineteenth-century Britain', in D. Fraser and A. Sutcliffe (eds) *The Pursuit of Urban History*, London: Edward Arnold.

Davis, L. (1995) *Enforcing Normalcy: Disability, Deafness and the Body*, London: Verso.

Dear, M. J. and Wolch, J. (eds) (1987) *Landscapes of Despair: From Deinstitutionalization to Homelessness*, Princeton, NJ: Princeton University Press.

Dorn, M. L. (1994) 'Disability as spatial dissidence: a cultural geography of the stigmatized body', unpublished MS thesis, The Pennsylvania State University, PA.

—— (1998) 'Beyond nomadism: the travel narratives of a "cripple"', in H. Nast and S. Pile (eds) *Places Through the Body*, New York: Routledge.

Drake, D. (1807) 'Observations on debating societies and the duties of their members; read before the Cincinnati Lyceum, 1807', unpublished manuscript in the Torrence Papers, Box 5, Cincinnati Historical Society, Cincinnati Museum Center, Cincinnati, Ohio.

—— (1814) Letter to Elizabeth Mansfield, from Cincinnati, Ohio to New Haven, Connecticut, in the Jared Mansfield Papers, Ohio Historical Society, Columbus, Ohio.

—— (1815) *Natural and Statistical View, or Picture of Cincinnati and the Miami Country*, Cincinnati: Looker and Wallace.

—— (1828) *A Discourse on Intemperance*, Cincinnati: Looker and Reynolds.

—— (1830) 'Medical Jurisprudence – Report of a trial for Murder, in which the culprit was defended on the ground of his labouring under Mania à potu, or Delirium from Intemperance', *The Western Journal of Medical and Physical Sciences* III, 1: 44–65.

—— (1841) 'Breweries vs. Foundries; Address to the Washington Temperance Society, Louisville, KY, December 25, 1841; and Chillicothe, July 4, 1842', unpublished manuscript in the Daniel Drake Collection, Cincinnati Historical Society, Cincinnati Museum Center, Cincinnati, Ohio.

—— (1842a) *Proceedings of the Physiological Temperance Society of the Medical Institute of Louisville*, Louisville, KY: N. H. White, Drake Collection, Department of Special Collections and Archives, University of Kentucky libraries, Lexington, KY.

—— (1842b) 'Travelling editorials: progress of total abstinence', *Western Journal of Medicine and Surgery* 5: 394.

—— (1847) *Strictures on Some of the Defects and Infirmities of Intellectual and Moral Character, in Students of Medicine: An Introductory Lecture, Delivered in the University of Louisville, November 1st, 1847*, Louisville: Prentice and Weissinger.

—— (1850a) 'Communication from Dr. Drake to the Medico-Chirurgical Society of Cincinnati', *The Western Lancet* XI, 9: 557–69.

—— (1850b) *A Systematic Treatise, Historical, Etiological, and Practical, on the Principal Diseases of the Interior Valley of North America, as They Appear in the Caucasian, African, Indian, and Esquimaux Varieties of its Population*, Cincinnati: Winthrop B. Smith & Co.

—— (1851) 'Causes and consequences of temperate drinking', in J. Young (ed.) *The Lights of Temperance*, Louisville, KY: Morton & Griswold.

Driver, F. (1988) 'Moral geographies: social science and the urban environment in mid-nineteenth century England', *Transactions of the Institute of British Geographers* 13, 3: 275–87.

Figlio, K. M. (1976) 'The metaphor of organization: an historiographical perspective on the bio-medical sciences of the early nineteenth century', *History of Science* 14: 17–53.

Foucault, M. (1972) *The Archaeology of Knowledge and the Discourse of Language*, New York: Pantheon.

—— (1974) *The Birth of the Clinic: An Archeology of Medical Perception*, London: Tavistock.

—— (1979) *Discipline and Punish: The Birth of the Prison*, New York: Vintage.

—— (1980) 'The politics of health in the eighteenth century', in C. Gordon (ed.) *Power/Knowledge: Selected Interviews and Other Writings*, New York: Pantheon.

—— (1985) *The Use of Pleasure – Volume 2 of The History of Sexuality*, New York: Vintage Books.

Frank, J. P. (1976) *A System of Complete Medical Police*, Baltimore, MD: The Johns Hopkins University Press.

Franklin, B. (1987) 'Death of a drunk', in J. A. L. Lemay (ed.) *Benjamin Franklin: Writings*, New York: Library of America.

Gleeson, B. J. (1996) 'A geography for disabled people?', *Transactions of the Institute of British Geographers* 21: 387–96.

Golledge, R. G. (1993) 'Geography and the disabled: a survey with special reference to vision impaired and blind populations', *Transactions of the Institute of British Geographers* 18: 63–85.

—— (1997) 'On reassembling one's life: overcoming disability in the academic environment', *Environment and Planning D: Society and Space* 15: 391–409.

Gomberg, E. S. L. (1982) 'Historical and political perspective: women and drug use', *Journal of Social Issues* 38, 2: 9–23.

Haller, J. S. (1981) *American Medicine in Transition, 1840–1910*, Urbana and Chicago, IL: University of Illinois Press.

Hayward, G. (1822) 'Some remarks on Delirium Vigilans; commonly called "Delirium Tremens," "Mania à Potu" or "Mania à Temulentia"', *New England Journal of Medicine and Surgery* XI: 235–43.

Helzer, J. E., Burnam, A. and McEvoy, L. T. (1991) 'Alcohol abuse and dependence', in L. N. Robins and D. A. Regier (eds) *Psychiatric Disorders in America: The Epidemiological Catchment Area Study*, New York: The Free Press.

Hippocrates (1983) 'Epidemics', in *Hippocratic Writings*, London: Penguin Classics.

Illich, I., Zola, I. K., McKnight, J., Caplan, J. and Shaiken, H. (eds) (1977) *Disabling Professions*, London: Marion Boyars.

Imrie, R. (1996) 'Ableist geographers, disablist spaces: towards a reconstruction of Golledge's "Geography and the disabled"', *Transactions of the Institute of British Geographers* 21, 2: 387–96.

Jarvis, E. (1850) 'The influence of distance from and proximity to an insane hospital on its use by any people', *Boston Medical and Surgical Journal* 42: 209–22.

—— (1971) *Insanity and Idiocy in Massachusetts: Report of the Commission on Lunacy, 1855*, Cambridge, MA: Harvard University Press.

Johnson, L. D. (1843) *Martha Washingtonianism, or A History of the Ladies' Temperance Benevolent Societies*, New York: Saxon and Miles.

Jordanova, L. J. (1979) 'Earth science and environmental medicine: the synthesis of the late Enlightenment', in L. J. Jordanova and R. S. Porter (eds) *Images of the Earth: Essays in the History of the Environmental Sciences*, Chalfont St Giles, Bucks., UK: British Society for the History of Science.

La Roche, R. (1830) 'Spontaneous combustion', *North American Medical and Surgical Journal* 10, 19: 181–2.

Lender, M. E. (1973) 'Drunkenness as an offense in early New England: a study of "puritan" attitudes', *Quarterly Journal of Studies of Alcohol* 34: 353–66.

Marcus, A. I. (1991) *Plague of Strangers: Social Groups and the Origins of City Services in Cincinnati*, Columbus: Ohio State University Press.

McCormack, D. (1998) 'Spilling over/into sober streets: "the drunkenness of things becoming various"', paper presented at the Annual Meeting of the Association of American Geographers, Boston, MA, 27 March.

Metcalf, S. D. (1998) 'Attention deficits', *Linguafranca: The Review of Academic Life* 8, 2: 60–4.

Mitchell, D. T. and Snyder, S. L. (eds) (1997) *The Body and Physical Difference: Discourses of Disability*, Ann Arbor: University of Michigan Press.

Nadelhaft, J. (1987) 'Wife torture: a known phenomenon in nineteenth-century America', *Journal of American Culture* 10, 3: 39–59.

Parr, H. and Philo, C. (1995) 'Mapping "mad" identities', in S. Pile and N. Thrift (eds) *Mapping the Subject: Geographies of Cultural Transformation*, New York: Routledge.

—— (1996) *A Forbidding Fortress of Locks, Bars and Padded Cells: The Locational History of Mental Health Care in Nottingham*, Historical Geography Research Series No. 32: Historical Geography Research Group.

Philo, C. (1987) '"Enough to drive one mad": the organization of space in 19th-century lunatic asylums', in J. Wolch and M. J. Dear (eds) *The Power of Geography: How Territory Shapes Social Life*, London: Unwin Hyman.

—— (1995) 'Journey to asylum: a medical-geographical idea in historical context', *Journal of Historical Geography* 21, 2: 148–68.

Pickstone, J. V. (1994) 'Museological science: the place of the analytical/comparative in nineteenth-century science, technology and medicine', *History of Science* 32: 111–38.

Pile, S. (1996) *The Body and the City: Psychoanalysis, Space and Subjectivity*, London: Routledge.

Poovey, M. (1995) *Making a Social Body: British Cultural Formation, 1830–1864*, Chicago: University of Chicago Press.

Rabinow, P. (1989) *French Modern: Norms and Forms of the Social Environment*, Cambridge, MA: The MIT Press.

Radford, J. P. (1991) 'Sterilization versus segregation: control of the 'feebleminded,' 1900–1938', *Social Science and Medicine* 33, 4: 449–58.

Ray, I. (1838) *A Treatise on the Medical Jurisprudence of Insanity*, Boston: C. Little and J. Brown.

Rohrer, J. R. (1985) 'Battling the master vice: the evangelical war against intemperance in Ohio, 1800–1832', unpublished MA thesis, Kent State University, Kent, Ohio.

Roizen, R. (1997) 'How does the nation's 'alcohol problem' change from era to era? Stalking the social logic of problem-definition transformations since repeal', paper presented at 'Historical Perspectives on Alcohol and Drug Use in American Society, 1800–1997' conference, The Francis Clark Wood Institute for the History of Medicine of the College of Physicians of Philadelphia, 9–11 May. Available on http://www.roizen.com/rou/postrepeal.htm. Also forthcoming in S. Tracy and C. Acker (eds) *Altering the American Consciousness: Essays on the History of Alcohol and Drug Use in the United States, 1800–1997*, Amherst, MA: University of Massachusetts Press.

Rorabaugh, W. J. (1979) *The Alcoholic Republic: An American Tradition*, Oxford: Oxford University Press.

Rosen, G. (1974) *From Medical Police to Social Medicine: Essays in the History of Health Care*, New York: Science History Publications.

Ross, S. J. (1985) *Workers on the Edge: Work, Leisure, and Politics in Industrializing Cincinnati, 1788–1890*, New York: Cornell University Press.

Rush, B. (1934) 'An inquiry into the effects of ardent spirits upon the human body and mind with an account of the means of preventing and the remedies for curing them', in Y. Henderson (ed.) *A New Deal in Liquor: A Plea for Dilution*, Garden City, NJ: Doubleday, Doran & Company.

Sellers, C. (1991) *The Market Revolution*, New York: Oxford University Press.

Shapiro, H. D. and Miller, Z. L. (1970) *Physician to the West: Selected Writings of Daniel Drake on Science and Society*, Lexington: University of Kentucky Press.

Short, C. W. (1830) 'Case of spontaneous combustion of the human body', *Transylvania Journal of Medicine* 3, 1: 143–5.

Stallybrass, P. and White, A. (1986) *The Politics and Poetics of Transgression*, London: Methuen.

Starr, S. Z. (1982) 'Prosit!!!! A non-cosmic tour of the Cincinnati saloon', in D. H. Tolzmann (ed.) *Festschrift for the German-American Tricentennial Jubilee*, Cincinnati: Cincinnati Historical Society.

Stone, D. A. (1981) 'The definition and determination of disability in public programs', in G. L. Albrecht (ed.) *Cross National Rehabilitation Policies: A Sociological Perspective*, London: Sage.

—— (1984) *The Disabled State*, Philadelphia, PA: Temple University Press.

Szaraz, S. C. (1993) 'History, character, and prospects: Daniel Drake and the life of the mind in the Ohio Valley, 1785–1852', unpublished Ph.D. thesis, Harvard University, Cambridge, MA.

Thomson, R. G. (1997) *Extraordinary Bodies: Figuring Physical Disability in American Culture and Literature*, New York: Columbia University Press.

Tyrrell, I. R. (1982) 'Women and temperance in antebellum America, 1830–1860', *Civil War History* 28, 2: 128–52.

Vujakovic, P. and Matthews, M. H. (1994) 'Contorted, folded, torn: environmental values, cartographic representation, and the politics of disability', *Disability and Society* 9: 359–75.

Walker, R. A. (1981) 'A theory of suburbanization: capitalism and the construction of urban space in the United States', in M. J. Dear and A. J. Scott (eds) *Urbanization and Urban Planning in Capitalist Society*, London: Methuen.

Warner, J. H. (1986) *The Therapeutic Perspective: Medical Practice, Knowledge, and Identity in America, 1820–1885*, Cambridge, MA: Harvard University Press.

Zola, I. K. (1993) 'The sleeping giant in our midst: redefining "persons with disabilities"', in L. O. Gostin and H. A. Beyer (eds) *Implementing the Americans with Disabilities Act: Rights and Responsibilities of All Americans*, Baltimore, MD: Paul H. Brooks Publishing Company.

4

RHETORIC AND PLACE IN THE 'MENTAL DEFICIENCY' ASYLUM

Deborah Carter Park and John Radford

Introduction: statement of a problem

For the last century and a half, whatever has passed for public policy in the realm of intellectual impairment has been preoccupied with questions of social and spatial positioning. Where do people with intellectual impairments fit into society? Where should they live? Should they work, and if so, where? Is it worthwhile educating them, and if so where and how? Should they be allowed to have children? Are they capable of exercising proper control over their own lives, and if so, in what kinds of environment?

Up to the 1970s the answers to such questions tended to assume the desirability – if not always the practicability – of exclusion. Since then the answers have been more varied. Exclusion has persisted in some cases, but progressive forces began to insist on principles of 'normalisation': the provision of environments and lifestyles as 'normal' as possible, situated within the social mainstream. More recently, the emphasis has been on the rights of persons with an intellectual impairment to live as independently as possible, with the ability to gain access to all entitlements enjoyed by society in general.

The period between the 1850s and the 1970s was dominated by a search for 'a place for mental deficiency', away from the mainstream. This resulted in the adoption of the asylum as the most desirable policy instrument. While its role was continuously debated, the asylum became the established norm. At first the objective of the asylum was to provide education for children, especially those in financial need, within a closed environment removed from the stresses of poverty, overcrowding, deviancy and other social problems. Later this goal became obscured by other imperatives, and the asylums were transformed accordingly. As the asylum was put to a succession of different uses it remained the major locus of policy implementation, and what went on inside the asylum affected the treatment of those similarly stigmatised who, for a variety of reasons, were never institutionalised. Incarceration in an asylum became the ideal 'solution' for the 'treatment' of 'mental deficiency', and this 120-year period can justly be called the 'asylum era' in the history of intellectual impairment.

The competing policy objectives which characterised the asylum era reflected a confusion over the nature and treatment of intellectual disability. While elements of a humane approach were occasionally evident – especially in the early years – the era as a whole was characterised by discourses of 'defect', 'deviance' and 'threat'. Intellectual capacity was increasingly viewed as quantifiable, hierarchical and mainly hereditary. Much of the research on the nature of intelligence was conducted in close association with the asylum managers, who often used the residents as their research subjects. This produced a remarkable degree of circularity of discourse, evident in the close fit between the succession of uses for the asylum and the shifting lexicon of rhetoric, classification, diagnosis and therapy.

By the late 1960s, it became evident that the asylum era had peaked, this at a time when discussion of its role was reinvigorated. By the early 1970s, it was clear that the end was near for mental deficiency asylums, as had been evident for mental illness asylums for at least a decade. As a subject for study, the mental deficiency asylum has been given less attention than the insane asylum and related institutions. Our focus in this paper is on those asylums which came to be known most commonly in Britain as 'hospitals for the mentally handicapped', in the United States as 'training centres', and in Canada as 'regional centres'. These institutions trace a common origin in the mid-Victorian 'idiot asylum'.

The succession of uses for the asylum was accompanied by a constantly changing vocabulary of classification, diagnosis, and recommended treatments. We examine the creation of a place for mental deficiency and the link between public policy and the creation of the mental deficiency asylum during the era labelled 'modern'. Our principal focus is the interrelationship between institutional treatment models, discourse, geodemographic variables, related institutions and constructs. To understand this interrelationship, we explore the rhetoric of cost, professional expertise and sexuality in the context of nation, race, morality and the body. We argue that this complex interrelationship is best understood by examining primary documents.

Rhetoric and place in England, the United States and Canada

Several accounts (Heaton-Ward 1978; Scheerenberger 1983) have described a chronology of developments by which the earliest asylums came to be established. There is general agreement that the earliest motivations were paternalistic, characterised by an optimism that children diagnosed as 'idiots' could be educated if they were placed in the right environment. The goal – which was clearly influenced by trends in the treatment of mental illness – was to separate children into closed environments 'away from the mainstream' for a period of years to prepare them for an independent life in the outside world. Early ventures in Switzerland by Guggenbuhl and in France by Itard are credited with stimulating the foundation of small charitable residential institutions in England beginning in 1846. These were short-lived, but were replaced by more ambitious purpose-built charitable asylums, the prototype being the Earlswood asylum near Redhill, founded in 1855. By the 1870s a number of asylums were tapping into a system whereby fees were paid by parish ratepayers through the Poor Law guardians. The catchment areas expanded beyond the local region, in some cases becoming virtually national in extent (Radford and Tipper 1988).

The geography of the nineteenth century idiot asylums in England has been analysed by Philo (1987) who has tabulated the institutions by type and plotted their locations. Philo's research led him to question the prevailing view that the remote locations of these asylums were a reflection of an 'out of sight, out of mind' motive. He maintained that the reformist search for therapeutic environments was genuine, citing the preponderance of mid-century rhetoric which stressed this concern. Later in the nineteenth century, however, arguments became much more exclusionary in tone. Philo suggests that this is a result of several decades during which the desirability of separation from the community – while well-intentioned – had generated a sense that something alien lay buried in the 'mentally deficient'. Philo's point, that we should distinguish the motives that moulded the asylums from those that came to predominate in later years, especially after the rise of the eugenics movement, has great merit. Yet it is also possible to see in mid-nineteenth-century writings the seeds of what came in later years to be a full-blown discourse of threat and exclusion (Gelband 1979).

The rhetoric of reform, improvement and optimism was even more pronounced in mid-nineteenth-century United States than in England. This reflects the fact that the origins of the asylum in the United States were rooted even more explicitly in the Enlightenment project, both in their resonance generally within the culture and more specifically through the influence of Edouard Seguin. In 1876, Seguin became president of the Association of American Institutions for Idiotic and Feebleminded Persons, newly formed to give professional status to the field. From the six founding asylum managers the Association grew rapidly, reflecting the proliferation of specialised asylums in the latter decades of the century (Trent 1994: 67–9).

The early emphasis in the United States was clearly on the education of 'idiots' and the tone was progressive. As in England, however, it is difficult to gauge how much of this rhetoric is to be taken at face value. Take for example the following quotation:

Many a town is now paying an extra price for the support of a drivel-
ling idiot, who, if he had been properly trained, would be earning his
own livelihood, under the care of decent persons who would gladly
board and clothe him for the sake of the work he could do.

This observation was made to the Massachusetts Legislature by one of the most
ardent progressives of the era, Samuel G. Howe. It was contained in a report
of 1848, at the height of what is frequently regarded as the educational phase,
and it was offered to encourage the appropriation of public money for asylum
care. 'Idiots' were cast, even by this eminent reformer, as unproductive unless
subjected to asylum training. It was not long before a medical model was super-
imposed on the educational one. Creation of 'a place for idiocy' was capped by
medicalising the school and creating the asylum.

Developments in Canada were influenced by trends in both England and the
United States. Premised on similar discourses, developments in Canada lagged
behind those south of the border so that the early educational phase was almost
absent in Canada. Like the mental illness asylum, mental retardation policy was
a provincial responsibility, and institutional foundations and growth were developed and implemented at different junctions in time, on different scales and in
varying geographical locations. But unlike the mental illness asylum, Canadian
mental handicap asylums were placed beyond the reaches of large Canadian urban
centres.

The first and largest facility in Canada was established in 1876 in the town of
Orillia (Ontario), 60 miles north of Toronto. Originally known as the Orillia Idiot
Asylum, the facility was located on the northern shore of Lake Simcoe, about one
mile south of Orillia. Better known as the Huronia Regional Centre, Orillia was
the sole facility in Ontario until 1950, when a facility for men was opened in the
town of Aurora a year later and Smith's Falls was opened in the eastern region of
the province. The second Canadian facility established was the New Westminster
Asylum in 1877 in British Columbia. Originally designed for the mentally ill, the
facility accommodated what was then referred to as the 'idiot class'. In 1890
the Portage La Prairie institution in Manitoba was established and while this facility accommodated a mixed population, in 1930 it was denoted as a facility for the
care of the mentally retarded. Almost two decades later, another facility,
Essondale, was established in 1913 in British Columbia in response to the over-
crowded conditions at New Westminster. Another major institution in Western
Canada was located in Red Deer, Alberta; and established in 1923. It was reported
that the facility was located 'advantageously in Central Alberta, equidistant from
Edmonton and Calgary . . . [it] overlooks the City's urban areas and yet is adjacent to an attractive rural setting which provides a contrasting pastoral environment of benefit to the retardate' (ASH Report n.d.: 4). In 1958, adjacent to the

Red Deer facility, Deerhome was established for the custodial care of adult 'mental defectives'. Similar to the rhetoric of 'mental deficiency' expressed in England and the United States, the origin of the Canadian asylum was rooted in a shifting lexicon of classification, diagnosis and therapy.

Interpreting the trends: policy palimpsest

It is clear that the role of the mental deficiency asylum changed over time, fulfilling a number of roles ranging from education to protection to social welfare to custodialism. The literature has continued to promote such a conceptual framework, advancing unilinear stage models in which changes in asylum care are presented as a natural progression in models of treatment according to a sequence of periods (Scheerenberger 1983; Tyor and Bell 1984). Although useful for narrative historiography, this rather simplistic chronological approach fails to embody significant changes in policy in a way that allows for comparative analysis. An alternative approach would be the conceptualisation of a stage model applied to various geographical jurisdictions to document policy evolution and local implementation. The inherent danger of such models is their tendency to force analysis into an artificially contrived unilinear path. Clearly, such models fail to account for the multifarious discourses and ideologies characteristic of mental handicap policy environments.

Simmons (1982) presented a refined view of such changes which avoided the unilinear models assumption of progress by allowing for the coexistence of various imperatives in asylum treatment, care and development. Simmons entitled this conceptual framework the 'policy palimpsest', representing layers of institutional treatment models and not necessarily stages as had been the case with the unilinear stage models. This schema identifies specific aspects of the policy environment that tended to ebb and flow over time, while recognising the possibility of various mixes occurring at particular junctures. The policy palimpsest was a step forward in the conceptualisation of mental deficiency asylums, avoiding the previous tendency towards a unilinear, whiggish interpretation of treatment history.

Accommodated within the policy palimpsest are the imperatives of public policy, visible in the early and later years of asylum development. The palimpsest includes four contrasting institutional treatment models. These may be listed from most to least prominent, as the palimpsest is viewed from the perspective of the turn of the century. In Simmons's view, the dominant influence in North American mental handicap policy by 1900 was a custodial model of care. Beneath this, although still clearly visible, was a less hostile social welfare approach and beneath this again, less legible, was the remains of an asylum approach: the institutionalisation of the mentally deficient for their own protection. At the bottom, and almost obliterated by 1900, was the original inspiration for the early institutions – the education of mentally handicapped children.

One advantage of this schematic framework is that it draws upon British, American and Canadian experiences, incorporating a comparative analysis. It is a persistent theme in Simmons's account that the trends were international in scope, and influenced by opinion-makers on both sides of the Atlantic. On the other hand, the policy palimpsest does not allow for mid-nineteenth-century discourses of custodialism. Yet, one can find in the records in Britain, the United States and Canada expressions of the custodial argument juxtaposed with educational and social welfare imperatives. The policy palimpsest still adheres to a sequence which intensive research fails to validate. It also short-changes the medicalisation of disability, which is one of the most persistent trends. In sum, it oversimplifies a complex situation.

Changing the perspective

Our discussion so far has dealt with a selection of the literature on intellectual disability which focuses on the attempts of officials to define, classify and 'name' the varieties of intellectual disability. That relationship is essentially reflexive: labels attached to diagnoses structure the observations of officials in their treatment of individuals, thus reinforcing the stereotype. But this relationship takes place within a multidimensional social and cultural context. It seems vital to specify at least two missing dimensions.

First, we urge the necessity of situating the discourses of the shifting currents of the asylum era in a structural context, augmenting the temporal one suggested in the policy palimpsest. Second, we suggest the need to get beneath the generalities expressed in the public pronouncements which constitute the principal sources for the investigation of discourse. This can be accomplished by examining primary documentation located in provincial archives.

The structure of the asylum mentalité

While the policy palimpsest offers an improvement on the unilinear models, it still falls short of representing the multidimensional structure of the social forces which influence public policy. A first step in specifying these forces is to identify four clusters of interrelated factors:

1   Officialdom, comprised of a number of sources of diagnosis and treatment including charity, law, medicine, psychology and social work.
2   Geodemographic variables, including gender, age, ethnicity, disability severity, urban/rural discrepancies and duration of stay.
3   Types of related institutions acting as alternative referral points, and participating in a cross-flow of 'patients'; these include orphanages, houses of refuge, mental illness facilities, prisons, isolation hospitals, and children's aid societies.

4   Constructions of apparent intellectual disability, including idiot, imbecile, moron, feebleminded, mental deficiency and mental retardation.

These are represented in Figure 4.1, and may be dealt with in turn.

The forces resulting in the confinement of people in asylums and the varying roles played by the mental deficiency asylum may collectively be termed 'officialdom'. In Britain, the 'mentally deficient' were defined objects of charity since little public money was invested in such facilities until after 1914. More particularly, charity defined suitable candidates for the leading asylums in Britain. On the other side of the Atlantic, in the United States and Canada, law, medicine, psychology and social work delimited a population subject to confinement in mental deficiency asylums. But the target population varied according to a host of other variables cutting across charitable, legal, medical and psychological discourses.

Under the rubric of geodemographic variables – by no means exhaustive – we have itemised gender, age, ethnicity, disability severity, urban/rural origin of admission and duration of stay. Institutional development was marked with a gender imbalance: prior to the 1860s, males composed by far the largest

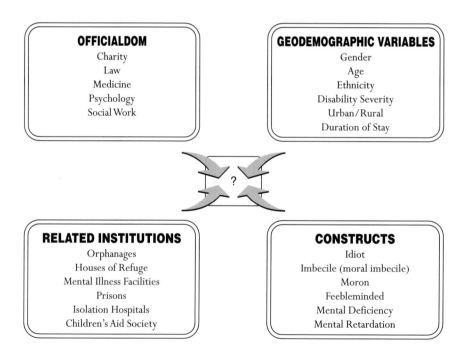

Figure 4.1 Constituent components in the evolution of the mental deficiency asylum

component of institutional populations, while after the 1860s, females made up the bulk of the population confined within the 'mental deficiency' asylum. A shift in gender balance also occurred in sterilisation activity in Canada and the United States in the early twentieth century (Reilly 1991). Age of admission to asylums also changed. Children were admitted with greater frequency than adults whereas in the earlier years of operation, during the height of the eugenics movement, individuals of child-bearing age came to be admitted to such asylums with greater frequency (Radford and Park 1993). In the case of ethnicity, in Western Canada, Ukrainians and native Canadians were more likely to be confined (Christian n.d.) and in the Orillia Asylum in Ontario, there was a greater preponderance of new immigrants and transfers from other institutions (Park 1995).

There was also a marked urban/rural imbalance in admission to the mental deficiency asylum, a much greater number of individuals being admitted from urban areas by the 1920s and 1930s – even when one allows for the urbanisation of the population in general (Radford and Park 1993). Duration of stay increased in line with the expanding custodial nature of the asylum. Also, before the turn of the century, the so-called 'idiot' preoccupied social reformers, while after the turn of the century, it was the mildly subnormal, or the 'moron', who were placed in such institutions with greater frequency.

Officialdom and geodemographic variables did not function in isolation but in relation to such institutions as orphanages, workhouses, mental illness asylums and other social welfare facilities such as those operated by the Children's Aid Society. The asylum itself played multiple roles. The linkages among these institutions are extremely complex and there is a need for detailed examination of rhetoric and praxis. The recent literature on deinstitutionalisation pays attention to transinstitutionalisation or 'decanting', but historically little work has been done on these interrelationships; yet it is clear that orphanages and houses of refuge were instrumental in the rise of the mental deficiency asylum.

In addition to the geodemographic variables and the related institutions one must assess the constructs of disabilities. Clearly all these are related; this is the point emphasised in Figure 4.1. Trent (1994) has focused on feeblemindedness as a construct, or, as he describes it, the invention of the feebleminded in the United States. Clearly we are not necessarily dealing with intellectual disability *per se* but with individuals who were targeted by legal and medical agents and institutionalised on the pretext of being idiotic, moronic, feebleminded, mentally deficient, mentally retarded and – as we will demonstrate – on the assumption of being socially and economically marginalised. We see a need to move away from the history of mental retardation to focus on the history of constructs and rhetoric; such an approach seeks to incorporate perception of disability as used in the framework of control.[1]

We therefore turn to primary documents located at the Orillia Asylum and at provincial archives during the asylum era so as to uncover how one particular asylum – the largest and earliest of its kind in Canada – was not only responding to local conditions but also mirroring similar developments in the United States and Britain. We examine how the mental deficiency asylum was responding to related institutions and the manner in which the rhetoric of dealing with this particular population gives one the impression that influences were coming from Britain but more particularly, the north-eastern United States. More significantly, we attempt to decode the rhetoric of cost, professional expertise and sexuality in the context of nation/race and individual/body.

In short, we are dealing with a problem seeking a definition or – more to the point – a problem which discovered a place but never quite found a definition. The places established for people labelled mentally deficient were places of uncertainty and their role was never really resolved. Such asylums became places to house a population for whom society had no other solution. It is this population which is represented in the primary documents.

Decoding the rhetoric: interpretation of archival documents

Institutional records such as annual reports, correspondence and case files of admitted individuals offer great potential for deepening our understanding of asylum rhetoric. These records not only allow insight into geodemographic variables, officialdom, related institutions and constructs of mental deficiency, but also illuminate how much of the discourse of disability was couched in a rhetoric of cost, managerialism and sexuality. Institutional case files, since they tend to document referrals, admissions and discharges, also permit the reconstruction of flows of people between institutions. They are especially good as sources for understanding officialdom. It should be obvious that these sources do not represent the voices of institutionalised residents. While it is recognised that alternative histories of the asylum such as those written from a patients' point of view add to our understanding, we hope to highlight how rhetoric and praxis were embedded in the discourses of professional experts and the individuals' judgements of officialdom.[2]

The nature of institutional archival documents is as diverse as are the approaches to institutional histories. There is a wealth of information pertaining to disability that can be researched in local and provincial archives. In this chapter we use provincial and institutional annual reports, surveys, meetings, minutes, case files and related correspondence. Where appropriate, the *Journal of Psycho-Asthenics* was consulted.[3] Evidence and analysis is drawn from archival documentation, in an attempt to explore the qualitative dimensions of mental deficiency, the asylum, rhetoric and praxis. Our goal is to use these sources to expose the complex relationships between officialdom, geodemographic variables, related institutions, constructs and rhetoric. This has the potential to get at the underlying rationales behind decisions on marginalisation, institutionalisation and institutional cross-flows. While we make no claim to represent individual patient experience and the nature of institutional life, we can begin to stand in the position of the patient and begin to build a picture from a patient standpoint. Some illustrations show how this can be done.

Rhetoric of cost and cost efficiency

A rhetoric of cost and cost efficiency, expressed by state and medical authorities, provided a structure to how mental deficiency was perceived. Fluctuating between a polarity of worth and worthlessness, structured by cultural, moral and social judgements and related to the social dangers of pauperism, delinquency, drunkenness and criminal activity, the 'mentally deficient' and 'feebleminded' posed a problem that was largely economic. They were 'unproductive' as a class, a disturbance to home conditions, and a burden to the public purse.[4]

By contrast, the institutionalisation and training of the 'feebleminded' was equated with solid fiscal management. In 1908, Helen MacMurchy, Ontario's newly appointed Inspector of the Feebleminded, justified a coherent policy towards feeblemindedness as follows:

> It is good business and good for the country. It would be still better for the country to make out of life's dregs and by-products, out of lives now wasted and worse than wasted . . . to make these persons something better.
>
> (MacMurchy 1908: 13)

One decade earlier a similar ethic was being expressed: 'The large increase in the number of working patients now employed is one of the best evidences of executive ability, professional skill and active oversight in institution management' (Christie 1897: xxiv). The asylum was widely seen as the vehicle to such progress; outside the asylum, the mentally disabled person was a liability. In the *Journal of Psycho-Asthenics*, an American-based periodical, the Secretary for the Indiana State Board of Charities commented that:

> the feeble-minded citizen does not amount to much as a laborer. He is more likely to be a dead weight upon his family or the community in which he lives. In many instances a single feeble-minded person proves to be such a burden that his whole family is kept in poverty and wretchedness in its effort to properly care for and support him. It is like putting

on brakes while the wagon goes up hill. In a poor asylum the feeble-minded inmate is of some use.

(Bicknell 1896: 51–2)

The records show numerous instances where mental deficiency and feeble-mindedness were cast in a rhetoric of social cost, citing a 'relationship between mental deficiency, drunkenness, criminality, pauperism, prostitution, and illegitimacy' (Parlby 1924: 3). The socio-cultural costs associated with the crimes committed by a 'feebleminded' population were clearly asserted in 1914 in a full page figure entitled 'What We Pay For The Feeble-minded in Ontario'. Reproduced

WHAT WE PAY FOR THE FEEBLE-MINDED IN ONTARIO	
1. For house rent, firing, furniture, food and clothes, medicine and medical attendance and nursing and care, while they do nothing for themselves.	Because they are maintained by night and day in hospitals, houses of refuge, refuges, orphanages, havens, rescue homes or by patriotic, benevolent, church and other societies, and municipalities and private persons, so that they shall not starve.
2. For the care of their children, largely illegitimate.	Because they have on average twice as many children as normal people.
3. For enforcement of law and order and the care of them in prisons, and the cost of trials and all other legal and judicial processes.	Because they are continually committing crimes over and over again.
4. For the consequences of their crimes.	Because they burn houses and barns, wreck railway trains, commit indecent assault and commit murder.
5. For the constant supervision of them in respectable homes.	Because it takes one good citizen's time to care for one mentally defective person in a home.
6. For the moral damage they do.	Because they harm and corrupt others with their evil ways, and are a temptation to others, and centres of immorality.
7. For the national and social unhappiness, degradation and deterioration they cause.	Because they bring into being unfit and foolish citizens, and thus pass on to coming generations the curse of Feeble-Mindedness.
8. For the loss of happy home-life, the great security of national and personal well-being.	Because the normal children are really deprived of the mother's care – she must give all her time

Figure 4.2 What we pay for the feeble-minded in Ontario

here as Figure 4.2, eight specific sets of costs, ranging from the 'national and social unhappiness, degradation and deterioration' to the 'consequences of their crime' to the 'moral damage they do', are documented. The feebleminded 'harm and corrupt others with their evil ways, and are a temptation to others, and centres of immorality'. When it came to criminal acts, they 'burn houses and barns, wreck railway trains, commit indecent assault and commit murder'. Even Walter Fernald, one of the more progressive commentators in the United States, could declare that 'Every imbecile is an incipient criminal' (Nova Scotia n.d.).

The Canadian National Committee for Mental Hygiene articulated the social costs of the mentally defective through a 4-D papier mache display at the Central Canada Exhibit held in Ottawa. Calculated to show the 'effects in the community of mental deficiency', the 4-D scenes were as follows:

> The first was a graphic representation of a mental defective's one-room home, showing the inevitable concomitant of pauperism resulting from the failure of society in providing the unfortunate individual in childhood with such training . . . The second scene showed two boys robbing a house – criminality being one of the common accompaniments of mental deficiency – while the third showed a woeful, striped-garmented figure behind the bars of a prison cell. The fourth scene was a replica of a section of a corridor in an old-time asylum showing two wretched figures seated on a bench waiting in deadening idleness for death to end their aimless and useless existences.
>
> (CNCFMH 1927: 2)

In their expression of concern, many activists invoked a narrative play between the polarities of inside and outside, light and dark, underscoring the extent to which social costs were spatialised. The stated intent behind the mental hygiene movement in Canada was to bring to 'light' the problems associated with the 'mentally deficient' and the 'consequent bringing out into the light of day of yet another grave social problem' (Parlby 1924: 4). 'Feeblemindedness' was widely regarded as having an impact on the social and economic fabric of a community. In Nova Scotia, for example:

> We may reasonably assume that this condition is responsible for a very considerable share of the pauperism, illegitimacy, vice and crime which exist in our Province, and we are aware that the defect is one which is singularly prone to be transmitted from parent to child. It would therefore, seem reasonable that from the economic, as well as from the moral and sociological points of view, a strong effort should be made to limit the multiplication of this unfortunate class.
>
> (Nova Scotia n.d.)

Patient case files echo this rhetoric. For instance, consider the case of an 11-year-old female admitted to Orillia in January 1942 from the Ottawa Children's Aid Society and discharged to an approved home in June 1963. The 'abstract of clinical history' reads:

> The Children's Aid Society have several times been notified that this woman and her husband went out drinking and neglecting the children who were small and sickly. The children are uncontrolled and used bad language. The children were always poorly dressed, dirty and verminous. This family has been known to five agencies since they moved into Ottawa in 1935 . . . CAS stated that they did not have enough foster homes to go around and that they could find no placement for this girl so in view of this fact and in view of the fact that it was difficult to learn at school, application was made at the Ontario Hospital School, Orillia.[5]

According to the 'mental status' report:

> Her only difficulty has been due to the fact that she has been telling the girls that there is no God . . . she can read and write but she cannot tell time. She doesn't know what capital means when you ask her what the capital of Ontario is. After a little explanation, she states that Ottawa is the capital of Ontario and that Churchill is the Premier of Ontario. She believes that there are ten provinces in Canada and says that she is now residing in Toronto.

Rhetoric of managerialism

A rhetoric of managerialism or professional expertise also pervaded the constructs of 'mental deficiency', expressed through the authority and power of science, medicine, education and social work. Standards of efficient administration, humanitarian care and scientific treatment were frequently mentioned in reports and correspondence, and permeated the delegation of authority to provincial, state and county departments of asylums, public charities and various agencies under the auspices of health. The authority invested in science, and particularly in medicine, was directly linked to officialdom.

In Ontario, until the 1960s and 1970s, authority for a feebleminded and mentally deficient population was delegated to provincial departments of asylums, prisons, public charities and health. All mental hospitals for the insane and feebleminded were regarded as health-related and administered by the province, usually the Provincial Secretary. In each facility there was a resident medical superintendent whose annual report was always couched within a rhetoric of professionalism and a discourse of surveillance and authority.

As early as the 1930s, however, institutions such as the Orillia facility felt the push to operate as both medical centres and training schools. Such institutions 'could no longer properly be described as asylums, but rather as Hospitals and research laboratories, and . . . this particular Institution is now a training school for the subnormal child' (McGhie and MacPhee 1929: 5). In 1928,

> all new patients entering the hospital are first received in the medical building where they are kept under observation for a month or two before being admitted to the general wards. Thus the danger of new patients bringing in contagious diseases is prevented, and an opportunity is afforded for the discovery and rectifying of any minor physical defects.
>
> (*Orillia Packet and Times* 1928)

As early as 1892, one of the Inspectors for Prisons, Asylums and Public Charities commented that a training and education function would provide Orillia with 'more of the character of a Hospital and Training school combined, than that of an asylum, and would no doubt add greatly to the reputation and popularity of the institution with the public' (*Schools* 1892). South of the border in 1896, the President of the Association of Psycho-Asthenics declared, that the asylum superintendent

> must cease to be a mere director. The wonderful discoveries of modern histology, biology and pathology, as portrayed in the literature of the past year alone, throw open an almost virgin territory to the scientific investigator . . . theory must go hand in hand with the microscope and scalpel.
>
> (Fort 1896: 26)

The rise of the specialised asylum was based on a rhetoric of expertise and the scientific study of mental deficiency. In the United States at an early stage of institutional development, the optimistic course of action was to

> enlist the interest of specialists entirely outside of institutions, and secure their services in such lines as appeal to them, and, for the love of science alone. Such institutions as are situated near large cities or strong medical institutions are to be especially envied for their opportunities in this direction.
>
> (*Journal of Psycho-Asthenics* 1896: 64)

And when it came to success in the scientific study of mental deficiency, 'there must be *first* of all talent and love for scientific study; *second*, ample time, not only for laboratory work, but for reading sufficiently to keep thoroughly informed as to the work of others along the same and similar lines' (*Journal of Psycho-Asthenics* 1896: 64). There was a well-recognised technique

> involving the scientific training of psychiatrists, psychologists and social workers . . . that technique in its scientific and experiential aspects is quite on a level with the technique of the medical profession in the diagnosis of physical disorders. It is recognised by the Law Courts, the Army, the Navy and the Air Force and it, in fact, is the main technique employed by the Directorate of Personnel of the Department of National Defence.
>
> (*Records of Eugenic Board* 1946)

Related to a rhetoric based on science was the medicalisation of mental disability and the active participation of key medical and political practitioners. In 1926, Dr John MacGregor of London, Ontario, commented that:

> of all the pursuits which occupy the minds of men, none should broaden the outlook more or give a more reasonable or lenient viewpoint than medicine. The knowledge it brings to us of ourselves and of the whole system of being about us, is a pursuit worthy of our most zealous and immutable effort.
>
> (*Immigration* 1926)

Almost thirty years earlier in the United States, on the occasion of the president's address to the Association of Psycho-Asthenics, the attending audience was requested to 'pause and consider where and how we can attain our proper position in the ranks of progress, and secure the influence rightfully our own'. Prior to the conclusion of Fort's address, he put forward a scheme for re-organisation stating that the Superintendent or Institution Head be a 'medical man' since

> Personally, I firmly believe that no such Institution should ever have a head officer not a medical man, for the very simple reason that men, however capable otherwise, cannot approach the subject of care, training and true scientific study of such defectives without preliminary training afforded by the medical school . . . as an Association we have to deal with facts, and these facts show us that our title is anomalous if non-medical Superintendents are admitted members.
>
> (Fort 1896: 30–1)

In the western province of Alberta, the Honourable Mrs Parlby confirmed before the United Farm Women of Alberta in 1924 that a 'medical man' should deal with this subject area, one 'who had been trained to deal with it from every angle, and who, through his own actual experience and research work amongst these people, could speak to you with authority' (Parlby 1924: 3). Almost a decade after the end of World War Two, at the Annual Meeting of the Advisory Committee on Mental Health in Ottawa, it was commented that there were some 'paediatricians and obstetricians telling parents that they had better not see the mentally defective baby and get rid of it as soon as possible' (*Papers on Advisory Committee* 1955).

Earlier in the century, mental hygiene advocates added preventative medicine to the managerial rhetoric, as demonstrated by the Canadian National Committee for Mental Hygiene:

> Mental hygiene is the last work in preventive medicine. Such types of failure as are represented in the asylum, the prison and poorhouse will, of course, always be with us; but mental hygiene is primarily addressed to preventing such failures wherever as possible.
>
> (CNCFMH 1923)

The recommendation of a mental hygiene approach was applied to mental deficiency and to psychiatry with the result that a Mental Health Program was approved by the provincial government in Ontario (*Annual Reports* 1918–1926, 1931).

Another dimension to this rhetoric of professional expertise was the role allocated to education and educational practitioners. In the Final Report of the Royal Commission on Mental Hygiene, it was reported that:

> it has been learned by experience that, if mental deficients are diagnosed when they are young, if they are given training suited to individual needs, and if they are supervised in the community when they leave school, the majority will give little trouble to society . . . The problem, therefore, is educational rather than medical.
>
> (*Commission's Final Report* 1928: G4)

On the other hand, a rhetoric based on education was intertwined with medical discourse. In one of MacMurchy's reports on the feebleminded, it was commented that the 'dregs' of the schools need to be 'rescued by the scientifically trained teacher, and the school physician, working under the kindly auspices of educational authorities who are too wise and progressive to neglect and despair the feebleminded child' (MacMurchy 1908: 13). This push for special classes was felt in other provinces. In Nova Scotia, it was reported that feebleminded individuals

> influence through suggestion or imitation the more unstable of the normal minded pupils to their degradation. In extreme cases they shock the pupils

of fine sensibilities, and frequently exercise a most pernicious influence upon the morals of other pupils. They absorb the time and attention of the teacher excessively as compared with the normal minded, and may dissipate the attention of the whole class, while the teacher is endeavouring uselessly to make a point understood by them. They distract not only the attention of the teacher, but of the whole school, thus breaking up the general attention to work, causing loss of time, temper and efficiency in the discipline of the school. If one of them in the school room gets some benefit, it may be at the expense of greater loss to perhaps thirty or even fifty.

(Nova Scotia n.d.)

The need for special classes for 'morons' and 'border-line' cases in all school systems was recommended by the Royal Commission on Mental Hygiene. Such classes were to

enable the individual to be fitted into some small niche in the industrial world where he can be self-supporting and be a contented citizen. It is from such that the world obtains most of its 'hewers of wood and drawers of water'.

(*Report of the Royal Commission* 1927: CC53)

Almost two decades earlier, special schools and classes were

an indispensable part of any scheme for the care and control of the Feeble-Minded, but it is only one part of such a scheme, and taken by itself it is about as useful as the work of a laundress who puts the soiled clothes to soak but never washes them, never wrings them out, never hangs them out to dry, never irons them, but just leaves them to soak.

(MacMurchy 1911: 18)

A rhetoric of managerialism was not necessarily restricted to science, medicine and education but found a home in the field of social work and legal matters; in many instances, the hard-core scientific rhetoric was intertwined with a rhetoric based in the social sciences. By the 1930s, in many instances, departments of social work were integral components in the identification and, in turn, probation and community placement of out-patients. From a very early stage, a link between mental deficiency and the legal system was expressed in various legislative acts of government. Consider the following:

admission of a mental defective to a school by order of a Judge is basically the same as admission of a mentally disordered person by order of a Judge to a mental hospital under the Mental Hospitals Act . . . Provision is made for a Judge to issue a warrant to apprehend a person alleged to be mentally defective and dangerous to be at large. Procedure for admitting such a person to a school is as above. Where there are no medical certificates for such a person the Judge need not release him, but may make an order for the custody of the person pending disposition of the case . . . The person making application for apprehension of an alleged mental defective must deliver to the Judge two medical certificates within five days after the appearance of the person before the Judge. If this requirement is not complied with, the Judge selects two physicians and orders them to examine the patient.

(University of British Columbia 1959: 7)

In the case of an 18-year-old female admitted to the Orillia Asylum in December 1940 by means of a Lieutenant Governor's Warrant, this young woman was charged with the murder of her infant son and was sent to Orillia for long-term care through a diagnosis of 'mental deficiency without psychosis'. As taken from the 'abstract of clinical history', it was reported:

she committed murder contrary to the criminal code and found unfit on account of insanity to stand trial . . . she was arraigned for trail on charge of murder and there was sufficient reason to doubt whether [Jane] was then, on account of insanity, capable of conducting her defence . . . jury was impanelled, evidence was adduced and the jury did return with a verdict that [Jane] was insane and unfit to stand trial . . . admitted to Orillia, Ontario Hospital until the Lieutenant Governor orders her back to imprisonment, if then liable thereto.

In 1927, based upon the findings of a royal commission in Canada, it was reported that the

investigation so far indicates that the problem involves our educational system as much as our social and public-health activities, but just where the line is to be drawn or to what extent these can be dovetailed into a complete and practical programme we are not yet prepared to recommend.

(*Report of the Royal Commission* 1927: CC23)

And in Alberta, where sterilisation was implemented legally from 1928 to 1972, 'confidence [was placed] in the highly trained and experienced psychiatrists, psychologists and social workers who are responsible for the selection of the cases for presentation, and the Eugenics Board, appointed as above indicated' (*Records of Eugenic Board* 1946).

Rhetoric of sexuality

The last rhetorical theme which we can identify stems from a particular notion of sexuality; a rhetoric of gender, reproduction, illegitimacy, prostitution and pauperism finding expression in programmes of institutionalisation and sterilisation. And related to the cost rhetoric already discussed, monetary value was inscribed within the rhetoric of sexuality.

Reproduction was central to institutional custodial care policy and the control of reproductive technology and sexuality. In Alberta, a programme of involuntary sterilisation was seen as not interfering with the natural right of the individual, since

> all rights, natural, social or legal, carry with them the responsibilities of using these intelligently and in the interests of society as a whole. Mental defectives are incapable for the most part of exercising self-discipline in sexual matters. If they marry, they are accordingly more reproductive than normal persons, and they are incapable of assuming the responsibilities for the proper physical care, education and moral training of their children.
>
> *(Records of Eugenic Board* 1946)

In many Canadian and American jurisdictions, officialdom's moral judgement was applied towards women of childbearing age, due to their assumed contribution to such problems as illegitimacy, pauperism and prostitution. Social commentators went to great lengths to show that a 'very large percentage of illegitimate children are the offspring of feeble-minded mothers'. In this rhetoric of reproduction, women were especially targeted by state and medical authorities: the 'feeble-minded woman falls easily into vice. She cannot in her weakness resist the persuasions and temptations which constantly beset her. With inherited abnormal passions, she must oppose not only the temptations which surround her, but her own dominating impulses as well' (Cummings 1902: 5). Dr Walter Fernald, the superintendent of the Massachusetts School for Feebleminded Children, stated:

> The tendency to lead dissolute lives is especially noticeable in the females. A feeble-minded girl is exposed as no other girl in the world is exposed. She has not sense enough to protect herself from the perils to which women are subjected. Often bright and attractive, if at large they either marry and bring forth in geometrical ratio a new generation of defectives and dependents, or become irresponsible sources of corruption and debauchery in the communities where they live. There is hardly a poorhouse in this land where there are not two or more feeble-minded women with from one to four illegitimate children each. There is every

reason in morality, humanity and public money, that these feeble-minded women should be under permanent and watchful guardianship.

(Bicknell 1896: 61)

In the same article, Bicknell – Secretary of the Indiana State Board of Charities – commented:

The immorality and demoralisation which thus often accompany the feeble-minded woman through life, leave in their train a harvest of illegitimacy and pauperism beyond the power of words to adequately portray. The three children of feeble-mindedness – Idiocy, Pauperism and Illegitimacy – are monstrosities from which we must protect ourselves. They are a triple burden upon the prosperity of the people and a threat against the best in morals and education.

(Bicknell 1896: 57)

The rhetoric of sexuality was persuasive in the language and images used to depict the alleged excessive sexual urges of the feebleminded. In Alberta, a woman convicted of being a 'common prostitute' was said to suffer from 'Nymphomania or excessive venereal impulse which is a symptom of her weak mindedness . . . her mental deficiency seemed to cause her to veer towards immorality' (*Mental Defectives Act* 1924). In the same province, the dissemination of lantern slides, posters and photographic images reinforced the negative stereotype of defectiveness. The circulation of such stereotypes was based upon the 'curse of vice and illegitimacy' associated with the 'evil of feeblemindedness' and in many instances, their alleged proclivity to reproduce rapidly was likened to 'abnormally developed animal passions' along with the lack of 'will and reason, which should control and repress them'. 'The feeble-minded woman . . . falls easily into vice. She cannot, in her weakness, resist the persuasions and temptations which beset her. When the baser passions are strong, she must oppose not only the influences from without, but her own dominating desires' (Bicknell 1896: 57).

At the federal level in Canada, a similar rhetoric was expressed in mental health posters. Value was placed upon 'stressing the importance to mental health of warm positive interpersonal relationships', accomplished by the following representation:

A scene of mother with a very young child showing an obviously warm and loving acceptance and handling of the child to again stress the fact that happy parents need happy babies who become happy children. This is an old theme but one that bears repeating and a striking photograph will freshen the message.

(*Papers on Advisory Committee* 1957)

One of the most poignant series of images was disseminated by the Department of Public Health in Alberta. Detailing a rhetoric of social purity and the threat of reproduction by a mentally defective population, this series of posters (circa 1920s) pushed home a particular narrative. In one image (Plate 4.1), the reader is shown a picture of two men with horses: one man was able to control his horse; the other is not so successful and has been thrown from the animal. Above the image appears the caption 'WHY IS IT NECESSARY TO CONTROL THE SEX IMPULSES?', and the message is clearly one which advocates the control of 'deviant sexuality'.

One final point worth mentioning has to do with the moral responsibility expressed by prominent women having a range of political affiliations with government. Framed within a rhetoric of their responsibility to the human race and nation, groups such as the Imperial Order Daughters of the Empire (IODE), The Women's Section of the Dominion Labour Party, United Farm Women of Alberta (UFWA) and the Women's Christian Temperance Union, along with such prominent women as Nellie McClung, Emily Murphy and Helen MacMurchy, frequently employed a rhetoric of cost, sexuality and reproduction in their published addresses and reports. These women were instrumental in the campaign to institutionalise and sterilise the mentally defective. In her address before the UFWA, the Honourable Irene Parlby closed with the following warning:

> I do think that as women, as mothers of the race, we should be considering this subject very seriously indeed. We should have every sympathy for those who are so unhappy as to have brought defective children into the world, through perhaps no fault of their own, except that of marrying into an unwholesome stock, a fact of which they may have been in entire ignorance, at the time of marriage . . . [As women we should] not allow sentiment to blink out common-sense . . . We can as women help in a very practical way at least in bringing up our own boys and girls to regard marriage as a very serious responsibility, not lightly to be undertaken without full knowledge of the kind of man or woman with whom the partnership is being entered into.
>
> (Parlby 1924: 11)

Conclusion: nation/race and individual/body

Connections between officialdom, geodemographic variables and related institutions have a long history in the construction of 'mental deficiency' and 'feeblemindedness'. Different aspects of this relationship were played out within a rhetoric of cost, managerialism and sexuality. A rhetoric of mental deficiency was manifested at two scales: nationalism/race and individual/body. For example, a rhetoric of cost and efficiency inscribed itself on the individual (in

WHY IS IT NECESSARY TO CONTROL THE SEX IMPULSES?

Nothing is more exhilarating than the experience of riding a spirited horse, but if uncontrolled he may unmount and kill his rider.

The sex impulse or drive is a noble gift of Nature if properly guided and controlled by Man's higher powers. Only when it is allowed to run wild does it become a base passion and a dangerous thing.

Plate 4.1 Why is it necessary to control the sex impulses?
Source: Provincial Archives of Alberta

the form of institutionalisation and the employment of patient labour in the asylum) and in the national consciousness (in the form of immigration policy).

Cost, managerialism and sexuality were framed within a rhetoric about the future of the race and nation. In many instances reference is made to immigration policy and the assumption that 'immigration is brought at too great a price if it entails the admission of any considerable number of individuals who will add to the burden of the nation caused by mental subnormality' (*Report of the Royal Commission* 1927: CC30). In the 1920s, the intolerance of 'others' piqued the imagination of the Canadian Federal government. In its report of 1921, the Canadian National Committee for Mental Hygiene stated under the sub-heading 'Effect of Immigration on the Problem of Mental Abnormality in Alberta', that it

> should be the aim of the Federal authorities to so guard our ports of entry that we do not receive an undue proportion of those who will become eventually a burden to the State. It is particularly desirable to reject the insane and mentally deficient because they often prove a greater menace than any other group.
>
> (*National Committee for Mental Hygiene* 1921: 42)

In another report, it was observed that

> Immigration is a crying need in this country. Our expansive fertile fields invite the coming of hundreds of men and women into this land of promise, but we must be very particular regarding the types that we admit. Unfortunately no small percentage of these who find their way here at the present time, and for some time past, have been of a definitely inferior type. Many of them are mental defectives of varying degrees who, shortly after coming, find their way into our hospitals and other public institutions . . . The medical profession can perform a lasting public service by bringing the matter to the attention of the Immigration Department . . . Is it not the time that some arrangement for permanently taking care of this group be instituted, in order that they, by some means, may be prevented from propagating their own kind? The medical profession can here step in and help set the machinery in motion which will safeguard future generations.
>
> (*Commission Correspondence* 1925–6)

Strict control of immigration advanced not only the 'national welfare' and 'well-being' but was an act of Christian citizenship. In her 'Report of the feeble-minded in Ontario' for 1910, MacMurchy commented:

Among ourselves, among our English-speaking kindred about the Seven Seas, and over the international boundary line of this continent, and in almost all civilised countries, we have already state institutions where devoted men and women help to fulfil the duties of Christian citizenship by parental care of just these weak and helpless children.

(MacMurchy 1910: 49)

MacMurchy asked that people listen to the 'silent cry' of the wronged feeble-minded woman:

The exceeding bitter cry of the wronged Feeble-Minded woman is a silent cry. She does not know how great the sin against her is. We do. She cannot protect herself against her great wrongs. We can. And if we do it not, at our hands will God require it.

(MacMurchy 1911: 63)

In the United States, a sense of nation and well-being found a place in a similar rhetoric couched in terms of race degeneration: 'Strong, intelligent, useful families are becoming smaller and smaller. Irresponsible, diseased, defective parents, on the other hand, do not limit their families correspondingly. There can be but one result. That result is race degeneration' (Human Betterment Foundation n.d.: 1).

This rhetoric of cost, managerialism and sexuality allows us to map the evolution of a polarity between the 'total institution' and the individual/body. 'Mental deficiency' was spatialised in buildings constructed for the purpose of custody and training. 'There is only one thing to do for idiots, imbeciles, and the very low-grade morons, and that is to place them in institutions where they can live out their helpless, hopeless existence in reasonable bodily comfort' (*Report of the Royal Commission* 1927: CC21). The training principle was expressed at the World's Fair in 1896:

Some idiots can be made self-supporting, just as an animal can be. I can take a horse and make it earn money, but it cannot earn money for itself. I can make an idiot earn money for me but he cannot earn it for himself. I can protect that idiot, as I can a child, but the idiot cannot protect himself . . . For this reason idiots have to be collected together, under the charge of trained attendants, and held for life.

(Bicknell 1896: 52)

Control of the 'mentally deficient' was so absolute that these institutions performed 'all those social obligations usually assumed by the home, school, church, etc.' (McGhie and MacPhee 1929: 9). At Orillia, 'few of the inmates have any direct contact with the outside world. They have ceased to be citizens and have lost the ordinary privileges of citizens, and unless they have faithful personal friends no one is interested in them individually' (*Toronto Globe and Mail* 1937).

A rhetoric of cost, managerialism and sexuality is also reducible to the 'mentally deficient' body. One of the central ideas in the writings of Foucault, is the notion that the body is spatialised as a controlling site. Segregation and sterilisation programmes were premised on control of the 'diseased' and 'defective' body. One of the first superintendents of Orillia commented:

> To produce improvement we must act on the whole being, body as well as mind; for idiocy is not simply a mental deficiency to be improved by a course of instruction directed to the mental faculties, but is also a deficiency of nervous functional power, the result of nervous or brain disease and largely depending on a feeble constitution.
>
> (*Schools* 1892: 4)

On the relationship between environment and body, another writer commented:

> In the poor asylum or private family he does not fit his surroundings. He is a round bolt in a square hole. In the large special institution the surroundings are made to fit him snugly and pleasantly. Being happier and more contented thus, he is more easily controlled and can and will do more and better work.
>
> (Bicknell 1896: 55)

Clearly, once institutionalised, few people escaped the 'total institution'. While travelling from 'coast to coast in Canada, visiting the mental hospitals en route', one observer commented: 'one is repeatedly reminded of one of H. G. Wells' novels; of the scenes in which is laid out a world that, according to the novelist's fancy, co-exists with ours, but which has no appreciable contact with it – in effect, a world within a world' (Le Bourdais 1927: 4). It is this ideology of total exclusion which has made the struggle for 'normalising', integration, and human rights for people with an intellectual impairment so difficult.

Notes

1 Various terms implying intellectual and social impairment have been applied to individuals on the basis of scant and distorted evidence. Given the primacy of classification and labelling in the creation of constructs of intellectual impairment, it is impossible to discuss this topic without using historically specific terminology, however inappropriate and discriminatory it is today. 'Mental deficiency', 'mental defectiveness', 'feeblemindedness', 'mental retardation' and similar 'diagnoses' were an integral part

of the constructs and rhetoric which sanctioned institutionalisation and sterilisation measures.

2 In the last couple of years, geographers have attempted to examine patient history and lifestories through sources such as case files, works of fiction and art. Medical historians have been progressive in their attempts at understanding patient history, picking up on what Porter (1985) characterised as 'doing medical history from below'. The more recent work of de la Cour (1997), Reaume (1997) and Park and Radford (1998) have employed patient case files to further their understanding of asylums and sterilisation.

3 The *Journal of Psycho-Asthenics* was one of the first periodicals devoted to the 'feeble-minded' problem. As the President of the Association remarked in 1896, the organisation has 'under our eyes the entire number of feeble-minded and a large percentage of the epileptic class, who are receiving care and treatment for study and research' (Fort 1896: 31). The journal was known as the *Journal of Psycho-Asthenics* from 1896 to 1940, the *American Journal of Mental Deficiency* from 1940 to 1987 and the *American Journal of Mental Retardation* from 1988 to the present, reflecting contemporary labels. A similar trend occurred in the name and renaming of institutions in Canada, England and the United States.

4 In Nova Scotia, in the mid-1920s, it was reported: 'Quite apart from the unproductiveness of this class and the economic loss which their presence entails, there is to be considered such factors as the disturbance of home conditions which result from the diversion of parental care from other and possibly normal members of the family to the afflicted ones' (Nova Scotia n.d.).

5 The cases cited in this chapter are attached to specific patient records. In accordance with freedom of information legislation, case file numbers and all other individual identifiers have been removed from the text and bibliography. This is to preserve anonymity.

References

Secondary sources

Bicknell, E. P. (1896) 'Custodial care of the adult feeble-minded', *Journal of Psycho-Asthenics* 1, 2: 51–63.

Christian, T. (n.d.) *The Mentally Ill and Human Rights in Alberta: A Study of the Alberta Sexual Sterilisation Act*, Edmonton: University of Alberta, Faculty of Law.

Christie, R. (1897) 'Twenty-ninth annual report of the inspector of prisons and public charities upon the lunatic and idiot asylums', *Ontario Sessional Papers* 10.

de la Cour, L. (1997) '"She thinks this is the Queen's Castle": women patients' perceptions of an Ontario psychiatric hospital', *Health and Place* 3, 2: 131–41.

Fort, S. J. (1896) 'Address of the President', *Journal of Psycho-Asthenics* 1, 1: 23–33.

Gelband, H. S. (1979) 'Mental retardation and institutional treatment in nineteenth century England, 1845–1886', unpublished Ph.D. dissertation, University of Maryland.

Heaton-Ward, W. A. (1978) *Left Behind: A Study of Mental Handicap*, London: Woburn.

Journal of Psycho-Asthenics (1896) 'The relation of public institutions to scientific investigation' 1, 2: 64–7.

MacMurchy, H. (1908) 'Report of the feeble-minded in Ontario', *Ontario Sessional Papers* 58.

—— (1910) 'Report of the feeble-minded in Ontario', *Ontario Sessional Papers* 23.

—— (1911) 'Report of the feeble-minded in Ontario', *Ontario Sessional Papers* 23.

Nova Scotia, Province of (n.d.) 'Report respecting feeble minded in Nova Scotia', Submitted to Lieutenant Governor of Nova Scotia by the Provincial Secretary (Appendix 33).

Park, D. C. (1995) 'An imprisoned text: reading the Canadian mental handicap asylum', unpublished Ph.D. dissertation, York University.

Park, D. C. and Radford, J. P. (1998) 'From the case files: reconstructing a history of involuntary sterilisation', *Disability and Society* 13, 3: 317–42.

Philo, C. (1987) *Convenient Centres and Convenient Premises: The Historical Geography of England's Nineteenth-Century Idiot Asylums*, University of Hull: Department of Geography, Working Paper No. 3.

Porter, R. (1985) 'The patient's view: doing medical history from below', *Theory and Society* 14, 2: 175–98.

Radford, J. P. and Park, D. C. (1993) '"A Convenient Means of Riddance": Institutionalisation of people diagnosed as "mentally deficient" in Ontario, 1876–1934', *Health and Canadian Society* 14: 369–92.

Radford, J. P. and Tipper, A. (1988) *Starcross: Out of the Mainstream*, Toronto: G. Allan Roeher Institute.

Reaume, G. (1997) 'Accounts of abuse of patients at the Toronto Hospital for the Insane, 1883–1937', *Canadian Bulletin of Medical History* 14, 1: 65–106.

Reilly, P. (1991) *The Surgical Solution: A History of Involuntary Sterilisation in the United States*, Baltimore, MD: Johns Hopkins University Press.

Scheerenberger, R. C. (1983) *A History of Mental Retardation*, Baltimore, MD: Paul H. Brookes Publishing Company.

Simmons, H. G. (1982) *From Asylum to Welfare*, Toronto: National Institute on Mental Retardation.

Trent, J. W. (1994) *Inventing the Feeble Mind: A History of Mental Retardation in the United States*, Berkeley, CA: University of California Press.

Tyor, P. L. and Bell, L. V. (1984) *Caring for the Retarded in America: A History*, Westport, CT: Greenwood Press.

Manuscript sources

Annual Reports: Correspondence, 1918–1926 (May 1931) (Toronto: Provincial Archives of Ontario).

'ASH report, recommendations and site masterplan' (n.d.) (Edmonton: Provincial Archives of Alberta).

Canadian National Committee for Mental Hygiene (1923) *Bulletin of the CNCFMH*, 1, 4 (Victoria: Provincial Archives of British Columbia).

—— (1927) 'Graphically depict mental deficiency at Central Canada Exhibit', *Bulletin of the CNCFMH* 2, 8 (Victoria: Provincial Archives of British Columbia).

Commission Correspondence (1925–6) 'General' (Victoria: Provincial Archives of British Columbia).

Commission's Final Report (1928) 'Final report of the Royal Commission on mental hygiene' (Victoria: Provincial Archives of British Columbia).

Cummings, Emily (3 November 1902) 'Report by Emily Cummings to Stratton, Secretary of the Province on Degenerates in the USA', *Return of Reports* (Toronto: Provincial Archives of Ontario).

Howe, Samuel Gridley (1848) 'Report made to the Legislature of Massachusetts upon idiocy' (Boston: Coolidge and Wiley).

Human Betterment Foundation (n.d.) 'Human sterilisation' *Inspection of Ontario Hospitals, 1931–1960*, Pasadena, CA: Human Betterment Foundation (Toronto: Provincial Archives of Ontario).

Immigration (July 1926) 'Letter from Dr. John MacGregor, London, Ontario' (Victoria: Provincial Archives of British Columbia).

Le Bourdais, D. M. (1927) *Canadian National Committee for Mental Hygiene*, 'Impressions from coast to coast', *Bulletin of the CNCFMH* 2, 7: 4 (Victoria: Provincial Archives of British Columbia).

McGhie, B. T. (Medical Superintendent, Ontario Hospital, Orillia) and MacPhee, E. D. (Associate Professor of Psychology, University of Toronto) (February 1929) 'Training and research in a Hospital for Subnormals', *Ontario Hospital, Orillia Publications* 1, 1 (Toronto: Provincial Archives of Ontario).

Mental Defectives Act (May 1924) 'Letter to the attorney general about a mentally defective woman' (Edmonton: Provincial Archives of Alberta).

National Committee for Mental Hygiene (Fall 1921) 'Mental hygiene survey of the province of Alberta' (Edmonton: Provincial Archives of Alberta).

Orillia Packet and Times (1 March 1928) 'Revolutionary changes in educational system at the Ontario Hospital: academic and industrial training to be co-ordinated, many improvements made in buildings' (Toronto: Provincial Archives of Ontario).

Parlby, I. (January 1924) 'An address delivered by the Honourable Mrs. Irene Parlby before the UFWA' (Edmonton: Provincial Archives of Alberta).

Papers on Advisory Committee on Mental Health, 1948–1959 (February and March 1955) 'Eighth meeting of the advisory committee on mental health' (Victoria: Provincial Archives of British Columbia).

—— (1957) 'Eighth meeting of the advisory committee on mental health' (1957) Letter from Dr Gee, Director of Mental Health Services to Dr Roberts, Principal Medical Officer, Mental Health, Department of National Health and Welfare (Victoria: Provincial Archives of British Columbia).

Records of Eugenic Board Meetings, Alberta (1946) 'Letter to the Knights of Columbus from the Secretary of the Eugenics Board' (Edmonton: Provincial Archives of Alberta).

Report of the Royal Commission on Mental Hygiene (1927) (Victoria: Provincial Archives of British Columbia).

Schools (February 1892) 'Letter from Christie to Beaton' (Toronto: Provincial Archives of Ontario).

Toronto Globe and Mail (25 March 1937) 'A disturbing report' (Toronto: Provincial Archives of Ontario).

University of British Columbia, Faculty of Law (1959) 'A report on mental health legislation in British Columbia' (Victoria: Provincial Archives of British Columbia).

5

CAN TECHNOLOGY OVERCOME THE DISABLING CITY?

Brendan Gleeson

Introduction: technological cures

According to some influential observers, disability is a 'natural ill' that technology can supposedly cure (Oliver 1990). In this view, the development of ever more sophisticated (and increasingly computerised) adaptive technologies – in the form of aids, appliances, and accessible urban design – will eventually liberate ever greater numbers of disabled people from the social and economic constraints 'imposed' by their bodily impairments (e.g., Sokolowska *et al.* 1981). Gordon (1983: 235) extols this view with his claim that:

> In this age of electronics, computers, and other marvelous advances which enable us to place men [*sic*] on the moon, we have the ingredients to give the disabled a place on the earth.

Gordon goes on to report the existence of various bionic and cybernetic 'antidotes' to disability. He is convinced that these 'rehabilitating' forces will help disabled people to overcome the socio-economic 'handicaps' they face, and thereby assume a meaningful role in society. Other techno-enthusiasts have also proclaimed that new aids and inclusive designs will progressively 'correct' for physical impairments and thereby turn the disabled person into a 'normal' citizen or worker. Steventon, for instance, believes that 'Technology is . . . Providing the ability to manufacture devices which offset many handicaps' (cited in Roulstone 1993: 241). Scherer's recent study of technology and disability in the United States found that engineers often 'see assistive technologies and environmental modifications as . . . the primary solutions to the functional limitations of a physical disability' (Scherer 1993: 84). As one engineer put it:

> There is nothing wrong with disabled people that the proper environment can't fix . . . Technology can solve anything . . . the problem is to get people to use devices.
>
> (Scherer 1993: 84)

As is evident above, faith is placed in the enabling powers of an expansive techno-logical project – namely, the application of science and engineering to everyday life – that ranges in scale from the smallest adaptive device (e.g., a hearing aid) to entire urban design technologies (e.g., accessible design techniques and regulations).

Whilst many theorists and disability activists would dismiss such claims about the social potential of adaptive mechanisms as exaggerated, even absurd, it is obvious that there exists within certain key institutional settings, such as the design and building professions and the medical establishment, a rather uncrit-ical faith in the power of such technologies to overcome the 'limitations of disabilities'. This faith in technology is often reflected in laws, policies, institu-tional arrangements and social attitudes which privilege technological solutions to the problems faced by disabled people. As the disability activist and commen-tator, Sally French (1993: 46–7), has remarked, 'technological aids are a mixed blessing' and can become 'a burden' when they are promoted to the exclusion of socio-economic policies for disabled people. Thus, for a better justification of a socio-political approach to disability issues, it is important to develop a crit-ical appreciation of the limits of 'technological solutions'.

In this chapter, I critically examine the notion that technological innovation – especially the techniques of urban design – can overcome the social exclusion and poverty that many physically disabled people experience in Western societies. This materialist view of disability does not, however, imply a narrow economic reduc-tionism. Two key materialist disability theorists, Oliver (1990) and Abberley (1997), have distanced themselves from social models which reduce the origins of disablement to purely economic causes. In contrast to such reductionism, the his-torical–geographical perspective that I will develop offers a richer framework that stresses the significance of a variety of material practices and representations emerging from culture, the economy and the state. Indeed, the materialist disabil-ity account is broadly similar to the cultural materialism of Raymond Williams (e.g.,1980). For example, when Finkelstein and Stuart (1996) point to the 'dis-abling culture' of contemporary capitalism, they refer to an ensemble of materially evident relations and representations, including political economic systems. Davis (1995) elaborates the cultural materialist view, pointing out how disability is socially produced through two interdependent 'modalities': 'function' and 'appearance'. Hence, disability is characterised both by political economic margin-ality (and even exploitation) and by cultural devaluation: a set of oppressive, inter-locking socio-spatial conditions. Here I concentrate on specific types of environmental technologies that have been used to enhance the autonomy of disabled people within built environments, including barrier-free design, personal mobility aids and accessible transport systems. In making this argument, I will oppose the 'technological determinism' – the idea that technology dictates social development (Jary and Jary 1991) – that sometimes characterises analyses of disability in Western social sciences and public policy realms. As I will show, this determinism is evident in recent urban geographic analyses that view disability as a naturally given set of limitations which can be reduced (if not abolished) through environmental technological innovations. Drawing upon the historical–geographical materialism developed by Harvey (e.g., 1996), I will propose an alternative conception of disability that views technology in non-deterministic terms. In this view, disability is neither 'a state of mind' to be overcome through heroic struggle or cultural change, and nor is it biologically determined. Instead, a historical–geographical perspective casts disability as a socio-spatial construction of the impaired body that can vary over space and time (cf. Dorn 1994).

This chapter focuses on the discrimination experienced by disabled people in Western (i.e., advanced capitalist) cities. The evidence here is drawn from North American, British and Australasian urban contexts which, notwithstanding local politico-economic specificities, broadly reflect the challenges of city life that confront disabled people in all Western societies. Disabled people have certainly suffered discrimination in other, non-capitalist urban settings, such as the former Soviet Bloc nations. I argue, however, that the discriminatory features of Western societies for disabled people – especially those sourced in commodity labour markets – require a specific theoretical and empirical consideration.

The structure of this chapter is as follows. The first section clarifies the notion of disability to be used in this discussion. Following this, I outline the problems of inaccessibility and environmental exclusion in contemporary cities. Then critical appraisals are made of two approaches evident in the geography and disability studies literatures which both propose different technological 'solutions' to the problems facing disabled people in cities. The fourth section outlines an alternative, 'historical–geographical' explanation of urban disability discrimination. The paper concludes by reflecting on the general relationship between technology and disability in a 'post-industrial' age.

Defining disability

'Disability' is a term that has many different uses in various places and is therefore hard to define objectively. Disability may be used to refer to a considerable range of human differences – including those defined by age, health, physical and mental abilities, and even family status – that have been associated with some form of social restriction or material deprivation. Here, I adopt the more focused sense of the term which is now commonly used in the social sciences – here 'disability' refers to the social experiences of people with some form of impairment to a limb, organism or mechanism of the body (Oliver 1990).

Thus, the sense of disability used here encompasses impairments that have an organic basis, including those which manifest as physical and intellectual impairments. This discussion will not directly consider the question of mental illness, a specific set of health-related conditions that can be distinguished from physical disability. Moreover, the analysis will focus on the experiences of people with physical impairments, as this 'sub-group' of disabled people have most often been the focus of various technological strategies and adaptations that have sought to enhance mobility and social integration in cities.

The disabling city

The modern dwelling, as Le Corbusier (1946: 210) famously declared in 1923, must be a 'House-Machine'. Subsequently, a generation of modernist architects rallied to the cause of designing workplaces and homes which would function as 'machines for living' (Jencks 1973:12). As integrated collections of buildings, circulation systems, and other functional spaces, cities too are technological artefacts. Indeed, they are, arguably, one of humanity's most significant technological accomplishments (Carter 1983).[1] As has been well documented (e.g., Mumford 1966; Stewart 1952), the scale and complexity of cities increased dramatically with the Industrial Revolution. From the late eighteenth century in Britain, the dynamic interplay of several socio-spatial forces – notably the growth of markets, rapid urbanisation and technological change – began to produce the urban industrial landscapes which characterised Western nations until recent times (Hobsbawm 1968).

A range of social scientists now agree that the industrial city was not a machine designed for disabled people. Hahn puts it succinctly:

> the built environment is basically designed for the average human being, plus or minus half a standard deviation. From the perspective of a bell-shaped curve, persons with many types of disabilities that place them in the tails of the distribution are effectively isolated by their environments.
>
> (1986: 273)

Moreover,

> In terms of ease or comfort, most cities have been designed not merely for the nondisabled but for a physical ideal that few human beings can ever hope to approximate.
>
> (1986: 273)

The analyses of Golledge (1993), Hahn (1986), Imrie (1996), Vujakovic and Matthews (1994) have identified various features of industrial and contemporary cities – including physical design, institutional policies, and mobility systems – that have prevented disabled people from participating in the mainstreams of urban social life in a range of Western countries. These analysts agree that the physical layout of cities – including both macro land-use patterns and the internal design of buildings – discriminates against disabled people by not accounting for their mobility requirements. Practically speaking, this 'disability discrimination' takes the form of:

1 physical barriers to movement for disabled people, including broken surfaces on thoroughfares (streets, guttering, paving) which reduce or annul the effectiveness of mobility aids (e.g., wheelchairs, walking frames);
2 building architecture that excludes the entry of anyone unable to use stairs and hand-opened doors;
3 public transport modes which assume that passengers have a common level of ambulance;
4 public information (e.g., signage) presented in forms that assume a common level of visual and aural ability.

The above list is not exhaustive but does include the more commonly recognised discriminatory aspects of the built environments of contemporary Western cities. Even allowing for the distinctive physical forms, economies, cultures, and planning policies of Western cities, the international breadth of concern raised by disabled people concerning inaccessibility demonstrates that this is a pervasive feature of urban life (Campbell and Oliver 1996; Napolitano 1996).

Whilst there seems to be growing agreement amongst urban geographers that the contemporary city ignores the needs of most disabled people, there is a considerable divergence of views both on why this discriminatory urban form has emerged and concerning how this social problem can be overcome. On closer inspection, these divergences reveal deeper theoretical differences over the role of technology in socio-spatial change.

A legacy of machines or nature?

Why did industrial and contemporary cities assume forms and adopt functions that discriminate against disabled people? In urban geographic analysis, three broad answers to this question have been given, and these can be characterised under the following headings: natural limits; thoughtless design; and historical–geographical construction. In this sub-section I will discuss the first two explanations, showing that both reveal aspects of technological determinism. The next sub-section will consider the relationship between disability and technology from a historical–geographical perspective.

Natural limits: technical solutions

The recent influential work of Golledge (1990, 1991, 1993) addresses the inaccessibility of the contemporary Western city for disabled people. His approach suggests that disabled people's mobility and accessibility problems are primarily caused by the 'natural limitations' imposed by their impaired bodies. These natural limitations can, to some extent, be compensated for through the use of specialised urban technologies (e.g., motorised wheelchairs, tactual maps, moving ramps, sonic navigational aids, laser canes), thereby enhancing the ability of disabled people to undertake meaningful social and economic activities (Golledge 1993).

Golledge argues that disabled people inhabit 'distorted spaces' (1993: 64): in fact, he argues that there is a unique 'world of disability' (1993: 65) which corresponds to the constricted time–space prism of the disabled individual. (Vujakovic and Matthews (1994: 361) echo this socio-spatial view with their stress on the 'contorted, folded and torn' environmental knowledges of disabled people.) This perspective offers the appealing idea that disabled people are actually creators of their own space, in that they are seen to actively transform the general geographic landscape in their everyday lives and reproduce this as their own experiential world. The main force, however, for this transformation is the disabled, 'naturally limited' body itself, that encounters the physical objects and surfaces of urban landscapes as daunting barriers to meaningful social participation. The result of this encounter is the production of 'distorted spaces . . . which these populations must *endure*' (Golledge 1993: 64, emphasis added).

The impaired body is not seen as the sole reason for spatial 'distortion': Golledge acknowledges that society – in particular, urban society – compounds the problems faced by disabled people. Therefore, he argues at several points that the physical urban environment is structured in ways that magnify the distorting effect of disability, through careless design and signage, for example, which inhibit the access and mobility of disabled people.

This approach reflects, amongst other things, methodological individualism (Oliver 1990). Disabled people's so-called 'worlds of disability' are seen to have a primarily physiological origin, located in the functional deficiencies of the individual body rather than in social patterns. These deficiencies are exaggerated, but not caused, by the physical arrangement of urban space. At this point, the view embraces technological determinism, seeing technical aids as the only way to overcome the natural limits of the impaired body. Environmental modifications that seek to increase access for disabled people are explained as 'efforts to *compensate* for disability' (Golledge 1993: 64, emphasis added). In this view, disability is a set of physiologically given deficiencies, rather than socially created limitations, for which society seeks to compensate the individual through environmental design concessions.

Disability scholars have raised profound objections to approaches which view the social, economic and political problems faced by disabled people as the inevitable consequence of individual functional inadequacies in an uncaring society. These critics, such as Oliver (1990) and Abberley (1987, 1991) have developed alternative 'social constructionist' accounts that explain disability discrimination as the product of a social organisation 'which takes no or little account of people who have physical impairments and thus excludes them from mainstream activities' (Oliver 1990: 11). In this view, technology can be both oppressive and emancipatory, depending upon the social uses to which it is put.

Even amongst geographers and urbanists who do not share Golledge's perspective on disability and its economic and political context, there is evidence in their analyses of another, perhaps more subtle, form of technical determinism. This determinism is revealed in analyses which see the inaccessible contemporary city as the uncorrected legacy of a 'thoughtless', and therefore discriminatory, industrial society.

Thoughtless design

The social sciences have produced several examinations of the inaccessibility problem that confronts disabled people in contemporary cities. Many of these studies have emerged from social sciences outside urban geography, including planning (e.g., Bennett 1990; Imrie and Wells 1993a, 1993b), architecture (e.g., Kridler and Stewart 1992a, 1992b, 1992c), and political science (e.g., Hahn 1986). For example, Hahn's (1986) landmark essay on disability and the city represented an important early step in creating a socio-political critique of inaccessible urban space in geography. Hahn advocated strongly for a social constructionist perspective on disability, arguing that the solution to inaccessibility:

> must be found in laws and policies to change that milieu rather than in unrelenting efforts to improve the capacities of the disabled individual.
>
> (1986: 274)

These investigations, and the continuing socio-political agitation of disabled people, have sought to shift the focus of public debate on access away from the disabled body (and its supposed natural limits) towards the institutional policies and practices that shape urban environments in particular ways. The success of

104

the social constructionist critique has been reflected in new public urban policies on disability issues in a range of Western countries, centring on access
legislation and inclusive building codes (Imrie 1996). In the United States, for
example, the enactment of the Americans with Disabilities Act in 1990 saw environmental accessibility defined as a civil right for disabled citizens (Harrison and
Gilbert 1992).

However, a major feature of the access literature is its frequent tendency to
reduce the social oppression of disablement to a built environment problematic
(Glendinning 1991). Too often both the academic and policy literatures on access
exhibit an over-simplified spatial analysis, reflected in arguments that see the
arrangement of the built environment as the principal source of disablement. The
major problematical spatial arrangements identified include cluttered and broken
surfaces (Imrie and Wells 1993b), poor cartographic information (Fry 1988;
Vujakovic and Matthews 1994), inaccessible transport systems (Brail *et al.* 1976;
Gant 1992) and low-density urban form (Hahn 1986). In this view, space is reduced
to an inanimate configuration of material objects, and its dynamic social character
is disregarded or under-emphasised. The assumed phenomenal form of space, the
city, simply becomes a static diorama, freed both from the social structures and
institutions which shaped it, and the biographies of people who animate it.

The geographers Imrie and Wells (1993a, 1993b) have produced insightful
analyses that highlight the roles that political economic and institutional factors
can play in preventing the successful implementation of public access policies.
However, in one such analysis they suggest that the capitalist city is:

> characterised by a thoughtless lack of design and planning in public and
> private building.
>
> (Imrie and Wells 1993a: 213)

This remark seems to deny, or at least underplay, the possibility that physical
inaccessibility arises from a socio-structural origin. Implied here is a view of
inaccessibility as a geographic 'mishap', whose diffuse origins are merely the
unconsciously (and sometimes deliberately) discriminatory decisions of individual
actors (i.e., planners, architects, developers) in the land economy. The discriminatory design of capitalist cities appears then to be an environmental 'accident'
that the state must correct through accessibility legislation, rather than the observable form of deeper material and ideological structures of discrimination. (For
Imrie, this implied view of disability as a historical–geographical 'accident' is
corrected in a recent thoroughgoing analysis (Imrie 1996) which exposes the
logic of political economic and institutional interests that have undermined access
policy in British and United States cities.) The 'thoughtless design' theory is also
evident in the disability studies literature: Topliss and Gould (1981: 128), for

example, have characterised inaccessibility as primarily the outcome of 'thoughtless disregard for the needs of disabled people'.

Although at times implying that inaccessibility is a conspiracy of 'environmental apartheid' practised by the ablebodied, Hahn (1986) at other points in his important essay evokes a rather simplified faith in the 'power of design' (i.e., better building and transport technologies) to solve the environmental problems facing disabled people:

> the prospect of designing a city in which all residents – regardless of their bodily capacities – would be given an equal opportunity to seek a satisfactory life seems well within the reach – if not the grasp – of modern endeavour. The creation of an urban environment adapted to the needs of everyone . . . is an objective that communities must strive to fulfil and a concrete as well as a theoretical possibility that appears worthy of major effort. In fact, probably the principal obstacles to the attainment of this goal are the limitations of the imagination.
>
> (1986: 273)

The clear implication here is that the challenge of creating an inclusive city is primarily a technical problem that can be overcome through the use of enlightened rational design, an enterprise worthy of the 'modern endeavour' that animates Western science and technology. Obviously the design professions, politicians and the general public would have to be won over to this new ideal (at least to some extent). From there, however, the project essentially involves turning expanded expert imaginations to the task of producing inclusive environmental modifications. In this sense, Hahn echoes the earlier analysis by Brail *et al.* (1976) of transportation systems for disabled people in the United States. For these urban analysts, transportation planning was 'concerned directly with distribution of access', meaning that the problem of inaccessibility could be 'solved' through the rationalisation of transit services to produce a more efficient and fairer mobility system (1976: 161).

Not much thought is given in either analysis to the socio-spatial structures (e.g., the capitalist land economy, regimes of production) or to the institutions (e.g., professions, national, regional and local states) which condition the production of space in Western cities. Do not these structures and institutions embody powerful alternative logics of spatial organisation? If such bodies and logics can be implicated in the production of inaccessible cities, both now and in the past, why would they simply embrace the new enlightened rational design as a mode of spatial organisation?

The 'thoughtless design' approach is technologically determinist for its implied faith in environmental solutions to the problems of inaccessibility and immobility. The approach is also technocratic because it assumes that expert (i.e., professionalised) knowledge can identify and implement technological improvements to cities (i.e. better design and environmental aids) that will remove, or at least greatly reduce, the problem of inaccessibility. Hahn (1986), for example, displays great faith in the ability of 'governmental policy' to mould the city to rational ends. The main policy and legislative forms of such technological correctives are access laws and building codes applied through local regulation and enshrined in national legislation (Imrie 1996). Thus, the 'thoughtless design' advocates share the enthusiasm of 'natural limits' proponents for environmental modifications to buildings, access routes and transport systems, as well as the provision of sophisticated adaptive technologies (e.g., sonar guidance systems), though the two perspectives differ markedly on why these technologies are needed.

In contrast to the access perspectives discussed above, the historical–geographical approach rejects the idea that urban inaccessibility can be countered through strategies which rely solely or even principally upon technological innovations in environmental design. The historical–geographical view sees technology in non-determinate terms, both as a reflection of social relations, and as a powerful influence upon societal arrangements. Technological innovation thus is as much a product of cultural and political economic ideologies and institutions as it is a force directing societal change (Jary and Jary 1991).

A historical–geographical explanation

The rise of capitalism

Historical–geographical theorists (e.g., Harvey 1996; Smith 1984; Soja 1989), and other post-positivist geographers, have argued for a view of space as socially produced; a socio-spatial inter-relation which sees society and space as mutually constituting material-symbolic dynamics. In this view, *capitalist* social space arises from the territorialisation of, amongst other things, deep structural forces, such as commodity relations, that, in the process of materialisation, are themselves mediated by existing spatial patterns. Importantly, the historical–geographical view locates the origins of disablement in capitalist society at the unseen and dynamic structural level of socio-spatial transformation: an interplay of social and spatial change that has devalued the capacities of physically impaired people.

One historical example of this socio-spatial relation is the growth of commodity relations in late feudal society which slowly eroded the labour-power of impaired people. Feudal life was governed by, amongst other things, a localised economy that encouraged cooperation rather than competition. In such a system there was no structural imperative to discriminate for economic purposes between 'stronger' and 'weaker' forms of labour-power. Different bodily capacities were accepted and matched with appropriate forms of labour, based upon a variety of considerations, including cultural patterns and the specific nature of the local peasant economy. In feudal England, for example, physically impaired people worked in small handicraft trades, such as weaving and cobbling, that did not require much mobility (Gleeson 1993).

This is not to advance a misty-eyed view of feudalism, which was in many instances a brutal, frugal social order. Moreover, there was cultural discrimination against disabled people in many times and places – especially when issues of impairment were confused with the general antipathy towards leprosy (Braudel 1981). The point being made is that there was little, if any, material basis for disability discrimination in feudal society. Nearly all feudal trades and work activities were more amenable to physically impaired people than were the labour regimes of industrial capitalism. Localised handicraft work, for example, was easier to carry out than labour set to mechanised and standardised rhythms. In industrialism, there were a few specific work spaces for impaired people – as doormen, guards, street vendors, entertainers, domestic workers – but these were shadows of the more inclusive labour forms that had existed under feudalism.

As markets appeared and spread in Europe during the early modern era (particularly from around 1400) this basic socio-economic sub-structure of the peasant order began to change. The increasing social authority of the law of value meant the submission of peasant households to an abstract external force (market relations) which evaluated the worth of individual labour in terms of average productivity standards. From the first, this competitive, social evaluation of individual labour-power meant that 'slower', 'weaker' or more inflexible workers were devalued in terms of their potential for paid work (Mandel 1968). Physically impaired workers thus entered the first historical stage of capitalism handicapped by the devaluing logic of the law of value and competitive commodity relations.

Also, the combined growth of commodity relations and urbanisation meant that sites of production were themselves evolving (rapidly by the late eighteenth century). Production sites were consolidating – first as 'manufactories' and later, with steam-powered mechanisation, as factories – and were transforming into urban social spaces that were compelled by the logic of competition to seek the most productive forms of labour-power. The 'original handicap' which early commodity relations bestowed upon impaired people was crucial in setting a trajectory of change in both the social relations of production and their technical-spatial forms (e.g., factories) that progressively devalued disabled labour-power.

The twin processes of labour commodification and urbanisation resulted in the production of increasingly disabling environments in Britain and its colonies. As was observed earlier, the emergence of the industrial city in the late eighteenth century crystallised the socio-spatial exclusion of disabled people that had been slowly rising after the appearance of commodity relations in the late feudal era. As Oliver puts it:

Changes in the organisation of work from a rural based, cooperative system where individuals contributed what they could to the production process, to an urban, factory-based one organised around the individual wage labourer, had profound consequences.

(1990: 27–8)

Morris describes these profound consequences:

The operation of the labour market in the nineteenth-century effectively depressed handicapped people of all kinds to the bottom of the market.

(cited in Oliver 1990: 28)

One disabling feature of the industrial city was the new separation of home and work, a socio-spatial phenomenon which was all but absent in the feudal era. This disjuncture of home and work created a powerfully disabling friction in everyday life for physically impaired people. In addition, industrial workplaces were structured and used in ways that disabled 'uncompetitive' workers, including physically impaired people. The rise of mechanised forms of production introduced productivity standards that assumed a 'normal' (namely, usually male and non-impaired) worker's body and devalued all others. Of course, the 'brutalism' of industrial capitalism resulted in the continual impairment of many 'normal' bodies.

As Marx (1981) pointed out at the time (1860s), one result of these changes was the production of an 'incapable' stratum of labour, most of which was eventually allocated to a new institutional system of workhouses, hospitals, asylums, and (later) 'crippleages'. Industrialisation and urbanisation, he believed,

produced too great a section of the population which is . . . incapable of work, which owing to its situation is dependent on the exploitation of the labour of others or on kinds of work that can only count as such within a miserable mode of production.

(Marx 1981: 366)

Is 'environmental re-engineering' the answer?

The social and technological history of capitalism foregrounds the processes of commodification and spatial change that progressively devalued the labour-power of physically impaired people. Given this socio-spatial genesis of environmental exclusion, it is appropriate to question the key assumptions which have underpinned many analyses of inaccessibility in the contemporary city.

First, the historical–geographical view exposes the inadequacy of relying *solely* upon environmental modification as a strategy for eradicating the deeply embedded social discrimination facing disabled people. For instance, the discriminatory design of workplaces often appears to disabled people as an immediate source of their social exclusion. This is true in only a very immediate sense. The real source of disablement is the set of social forces that produce disabling workplaces and exclusionary technology. One such force is the commodity labour market, which, through the principle of competition, assumes that certain individuals (or *bodies*) will be rewarded and enabled by paid employment, whilst others are disabled as socially dependent. These twin assumptions guide the production of employment environments and production technologies that encourage access only by those (non-disabled) bodies which are economically valued by labour markets.

Access regulations and inclusive building codes, where they can be successfully applied, may improve the employment chances of some disabled people. However, better building standards and new modes of mobility, will not *on their own* revalue the labour-power of all physically impaired people. They will not guarantee economic security and social acceptance for disabled people. Such strategies can reduce the friction of everyday life for disabled people, and must be defended for this, but they will not solve the dynamic socio-spatial oppression of disablement. As Scherer explains of the United States situation:

> An unfortunate state of affairs currently exists in which we are making widespread environmental accommodations and are creating more and better technologies that minimize the functional impact of a disability, yet we often fail to provide . . . essential and basic opportunities for assimilation. While accessibility and enhanced functioning are important goals, more crucial are an individual's basic needs for security, autonomy, affiliation, accomplishment, intimacy and identity.
>
> (1993: 85)

Moreover, the resistance to access regulations and environmental modifications by public institutions and private development interests in Western countries attests to deep, and enduring, structures of disability discrimination which shape contemporary cities. Although most Western countries now have in place some form of building and planning legislation which attempts to counter the problem of inaccessibility, there is accumulating evidence to show that such policies are generally failing to reduce or prevent discriminatory urban design (Gleeson 1997; Imrie 1996). Access legislation is often opposed by urban development interests, and governments tend to be less than rigorous in its enforcement. Recently in the United States, powerful corporate interests have argued before the federal judiciary that the Americans with Disabilities Act, by requiring businesses to provide wheelchair access, is an unnecessary restriction upon private property rights, and therefore an infringement of the Fifth Amendment (Helvarg 1995).

In Britain, Imrie (1996) has shown how the Thatcher government during the 1980s progressively relaxed central controls on accessibility standards, and encouraged a mood of regulatory voluntarism amongst local authorities (which bear the primary responsibility for enforcing access codes). The author argues that many local authorities subsequently gave little policy priority and few resources to accessibility responsibilities. The national lethargy on access policy was attributed in part to the flourishing climate of local growth politics, and the consequent anxiety of individual councils that 'superfluous' building regulations would frighten away increasingly mobile development capital.

Like Britain, New Zealand has enacted accessibility legislation, in the form of amendments to its Building Act (1991), which aim to make that country's cities more accessible to disabled people. However, as I have shown elsewhere (Gleeson 1997), there is increasing evidence that New Zealand's access regulations are being successfully resisted by development capital. In one major New Zealand city, Dunedin, disability advocacy groups have argued that local government has failed to enforce the accessibility standards in the building legislation.

Reflecting the British experience portrayed by Imrie, Dunedin disability activists have recently argued that the city's local government has neglected its accessibility policy responsibilities by underresourcing its building standards inspectorate. In late 1993, after some press exposure of activists' complaints, the city council's building control manager publicly admitted that 'resources are being stretched', and added the familiar bureaucratic excuse for inaction by remarking that 'the requirements [of the Building Act] are being enforced as far as is reasonably practicable' (*Dunedin Star Midweek* 10 November 1993: 1). In a further admission this same officer acknowledged that the council had not required a particular commercial establishment to install a lift during a major refit, although the building legislation may have required this. He then attempted to reassure the city's disability community with the observation that 'this place can still be accessed by people with disabilities who are not confined to a wheelchair' (*Dunedin Star Midweek* 10 November 1993: 1), thereby demonstrating a highly selective notion of disability that conflicted with the inclusive aim of the legislation.

Clearly, as technical 'solutions' to disability discrimination, inclusive design techniques and access regulations have serious limitations. This, of course, exposes the historical–geographical perspective to the charge of political impracticality – if access regulation and environmental modification cannot solve disability discrimination in the city, what will? An obvious target for change is the social system through which the labour of individuals is valued (and devalued). A start might be made with the labour process, a set of social and technological

111

relationships that presently devalue the work potential of disabled people (Barnes 1992). Both the design of workplaces and the technologies used therein have been shaped by socio-cultural perceptions of what constitutes a 'normal' (i.e., productive) body. However, if technological change, and its application in work settings, is socially conditioned, surely new institutional arrangements could produce an inclusive labour process?

There are some recent and continuing examples of attempts to create such new institutional arrangements. There have been some encouraging examples in Australia, for example, where governments during the 1980s pursued a series of active interventions in employment markets in order to enhance the value of labour by disabled people. Sheltered work arrangements were discouraged and assistance given to disabled people in open employment settings, sometimes involving subsidies for technological changes that shaped inclusive labour processes (Gleeson 1998). In Germany, also, the state has sought to reshape labour markets in favour of disabled people; here, division two of the federal Severely Disabled Persons Act 1974 establishes an 'Obligation of Employers to Employ Severely Disabled Persons' (the number of reserved posts depending upon the size of the firm). Again, the German government has sought to redirect the current of technological change within private and public sector workplaces by funding the creation of inclusive labour processes. These examples of active labour market policy contrast with a 'rights-based' approach which merely guarantees the prerogative of work for disabled people without providing the institutional means for such a goal to be realised.

Conclusion: a post-industrial utopia?

I began this chapter with reference to the techno-enthusiasm that pervades much of the disability studies literature. I set out from there to appraise critically those perspectives which see technology as the solution to the urban problems that face disabled people in contemporary cities. My critique focused upon two views that have been influential in the social sciences and public policy realms: what I termed the 'natural limits' and the 'thoughtless design' perspectives. In different ways, both approaches are technologically determinist for inappropriately emphasising the role of technology both as an oppressor of disabled people and as a means for overcoming problems such as inaccessibility and socio-economic marginalisation.

Critics may respond that I have simply focused on the wrong technologies in my appraisal, and that more significant technical innovations in the organisation of work, consumption and communication will *really* liberate disabled people from environmental and social discrimination. I refer here to the enthusiasm voiced since the 1970s by many influential observers of technical change (e.g., Bell 1974; Toffler 1971) for the 'post-industrial' society – a social economy transformed by telecommunications advances, the rising significance of information as a commodity, and the shift to service industries, resulting in new, flexible types of work and a dispersed, even home-based, geography of production and consumption.

In the disability literature, there has been great enthusiasm for this 'post-industrial' future (Oliver 1990; Roulstone 1993). As Weinberg has enthused:

> microprocessor devices and computer designs extend the disabled person's sense of autonomy and self-reliance and enable him or her to be more active in society.
>
> (cited in Roulstone 1993: 242)

The advent of new, adaptive work technologies – especially those framed around computers – and the dispersion of service employment through the use of networked telecommunications are seen as the answers to the problem of urban inaccessibility. Put simply, in the post-industrial scenario, work will be offered to disabled people in an appropriate form and carried out in their homes, therefore reducing the need for travel in physically inaccessible cities. In the post-industrial city, 'mobility' will be achieved through 'telecommuting', or 'virtual communication', rather than through physical travel.

Several objections can be raised to this form of technological determinism. First, at a general level, urban geographers and other social scientists have already shown that the 'post-industrial' city has not eventuated, at least not in the form imagined by the concept's enthusiasts (Kraut 1989; Scott and Soja 1986). Work, though certainly more geographically dispersed, computer dependent and service-based in some economic sectors, has none the less largely remained spatially concentrated in industrial and commercial settings, including the factories, shopping centres and offices that continue to be inaccessible for many disabled people. There has been a proliferation of home-working in many Western cities, though this has often tended to be a bi-polar phenomenon focused upon the re-emergence of new, exploitative domestic labour regimes and, on the other hand, professionalised forms of individual contract-based employment (Castells 1985). Neither of these new home-work forms are readily available, or desirable, for most disabled people, who remain handicapped by poverty and lower education standards in most Western countries (Oliver 1990).

The post-industrial society literature has pointed to a potential social divide between the 'information rich' and the 'information poor' (Jones 1990; Murdock and Golding 1989). By reason of their long experience of discrimination, there is every reason to believe that disabled people have struggled to join the ranks of the former group. Some observers (e.g., Croxen 1982) argue that new work technologies often disadvantage disabled people by requiring high levels of education, technical skills, self-confidence, and, in many cases, physical dexterity. In 1990, Susan Hammerman, Secretary General for Rehabilitation International, was forced to observe that, in the European experience, technology had not solved the employment problems of disabled people:

> A Commission of European Communities review of unemployment among disabled people revealed rates two to three times greater than among other segments of the populace. *A new danger cited within the European Community is the declining rate of successful placement of disabled people in employment, a fact which is ironic in light of the wonders of new technology available to virtually eliminate functional limitations as a factor at the work site.* The challenge of meshing new technological developments within existing infrastructures and within existing programs and approaches towards the rehabilitation of disabled people is enormous.
>
> (cited in Klugman *et al.* 1991: 30, emphasis added)

As the historical–geographical perspective maintains, technological change is always embedded in socio-political change. The present 'post-industrial'/'post-modern society' in the West is as much characterised by corporate downsizing, firm 're-engineering', labour market deregulation, and growing socio-spatial polarisation (Harvey 1996; Rifkin 1995), as it is by new and empowering forms of information technology (IT). There is increasing evidence that these other socio-political changes have particularly disadvantaged disabled people (e.g., Glendinning 1991; Klugman *et al.* 1991).

Increasingly, many disability activists and commentators are questioning the benefits of technology for disabled people (e.g., Scherer 1993). As Roulstone (1993: 242) acutely observes of new work technologies:

> whilst evidence of wider employment opportunities and enhanced employment is available . . . little evidence exists to show that new technology is redefining the *notion of disability*.
>
> (original emphasis)

Indeed, disability discrimination is a complex socio-political phenomenon that is sourced in deep societal structures, such as political economic relations, cultural dispositions, and institutional practices – the same forces which shape and are re-shaped by technology. It is impossible to imagine, therefore, that one dynamic of socio-spatial change – technology – can radically transform a deeply embedded social relation, such as disability. Moreover, as Scherer argues, assistive technology can be used to create a form of independence for disabled people that actually compounds their social isolation. She observes, sadly, that 'we have become less interdependent on people and more dependent on technology', concluding that 'technology is letting us down' (Scherer 1993: 167).

Disabled people desire, and have struggled for, a valued social role which ensures that their specific needs for material welfare, cultural acceptance, and socio-spatial inclusion are met. As the historical–geographical perspective shows, this valued place cannot be secured simply through technological innovations of any sort, but must be won through a lasting transformation of the political-economic, institutional and cultural forces that shape our cities and societies. As the philosopher Iris Young (1990) has cogently argued, this change must win a deep commitment to social inclusion in our structures, institutions and personal lives. Only with such a deep transformation can non-disabling spaces and places be socially valued as important, and therefore socially conceived and produced.

Technology cannot 'cure' disability, it cannot alone 'rehabilitate' physical impairments, as these 'deficiencies', though real enough to individuals, are always socially defined and must therefore be experienced as elements of human social relations. One disability commentator, Wolff, has observed that 'technology can never replace the quality of interpersonal relationships' for disabled people (cited in French 1993: 46). As Abberley (1991) argues, disabled people want acceptance, not assimilation, which means social respect for their physical differences, their unique abilities and vulnerabilities.

Acknowledgement

An earlier version of this chapter was published as 'A place on earth: technology, space and disability' in *Journal of Urban Technology* 5, 1: 1–19.

Note

1   Doubtless Le Corbusier would have endorsed such a characterisation – he made reference, for example, to the 'mechanism of the city' (1971: 62).

References

Abberley, P. (1987) 'The concept of oppression and the development of a social theory of disability', *Disability, Handicap, and Society* 2, 1: 5–19.
—— (1991) *Disabled People – Three Theories of Disability*, Occasional Papers in Sociology, no. 10, Bristol: Department of Economics and Social Science, Bristol Polytechnic.
—— (1997) 'The spectre at the feast – disabled people and social theory', unpublished paper, copy obtained from author.
Barnes, C. (1992) 'Disability and employment', *Personnel Review* 21, 6: 55–73.
Bell, D. (1974) *The Coming of Post-Industrial Society: A Venture in Social Forecasting*, London: Heinemann.

Bennett, T. (1990) 'Planning and people with disabilities', in J. Montgomery and A. Thornley (eds) *Radical Planning Initiatives: New Directions for Planning in the 1990s*, Aldershot: Gower.

Brail, R., Hughes, J. and Arthur, C. (1976) *Transportation Services for the Disabled and Elderly*, New Brunswick, NJ: Center for Urban Policy and Research.

Braudel, F. (1981) *Civilization and Capitalism – Volume One: 15th–18th Century. The Structures of Everyday Life: The Limits of the Possible*, London: Collins.

Campbell, J. and Oliver, M. (1996) *Disability Politics: Understanding Our Past, Changing Our Future*, London: Routledge.

Carter, H. (1983) *An Introduction to Urban Historical Geography*, London: Edward Arnold.

Castells, M. (1985) 'High technology, economic restructuring, and the urban-regional process in the United States', in M. Castells (ed.) *High Technology, Space and Society*, Beverley Hills, CA: Sage.

Croxen, M. (1982) *Disability and Employment: Report to the Commission of European Communities*, Brussels: Commission of European Communities.

Davis, L. J. (1995) *Enforcing Normalcy: Disability, Deafness, and the Body*, London: Verso.

Dorn, M. (1994) 'Disability as spatial dissidence: a cultural geography of the stigmatized body', unpublished MSc thesis, The Pennsylvania State University.

Finkelstein, V. and Stuart, O. (1996) 'Developing new services', in G. Hales (ed.) *Beyond Disability: Toward an Enabling Society*, London: Sage.

French, S. (1993) 'What's so great about independence?', in J. Swain, V. Finkelstein, S. French and M. Oliver (eds) *Disabling Barriers – Enabling Environments*, London: Sage.

Fry, C. M. (1988) 'Maps for the disabled', *The Cartographic Journal* 28: 20–8.

Gant, R. (1992) 'Transport for the disabled', *Geography* 77, 1: 88–91.

Gleeson, B. J. (1993) 'Second nature? The socio-spatial production of disability', unpublished PhD thesis, University of Melbourne.

—— (1996) 'A geography for disabled people?', *Transactions of the Institute of British Geographers* 21, 2: 387–96.

—— (1997) 'The regulation of environmental accessibility in New Zealand', *International Planning Studies* 2, 3: 367–90.

—— (1998) 'Disability and poverty', in R. Fincher and J. Nieuwenhuysen (eds) *Australian Poverty: Then and Now,* Melbourne: Melbourne University Press.

Glendinning, C. (1991) 'Losing ground: social policy and disabled people in Great Britain, 1980–90', *Disability, Handicap and Society* 6, 1: 3–19.

Golledge, R. G. (1990) 'Special populations in contemporary urban regions', in J. F. Hart (ed.) *Our Changing Cities*, Baltimore, MD: Johns Hopkins University Press.

—— (1991) 'Tactual strip maps as navigational aids', *Journal of Visual Impairment and Blindness* 85, 7: 296–301.

—— (1993) 'Geography and the disabled: a survey with special reference to vision impaired and blind populations', *Transactions of the Institute of British Geographers* 18, 1: 63–85.

Gordon, E. (1983) 'Epithets and attitudes', *Archives of Physical Medicine Rehabilitation* 64: 234–5.

Hahn, H. (1986) 'Disability and the urban environment: a perspective on Los Angeles', *Environment and Planning D: Society and Space* 4: 273–88.

Harrison, M. and Gilbert, S. (eds) (1992) *The Americans with Disabilities Handbook*, Beverley Hills, CA: Excellent Books.

Harvey, D. (1996) *Justice, Nature and the Geography of Difference*, Oxford: Blackwell.

Helvarg, D. (1995) 'Legal assault on the environment', *The Nation* 30 January: 126–7.

Hobsbawm, E. (1968) *Industry and Empire: An Economic History of Britain Since 1750*, London: Weidenfeld and Nicolson.

Imrie, R. F. (1996) *Disability and the City: International Perspectives*, London: Paul Chapman.

Imrie, R. F. and Wells, P. E. (1993a) 'Disablism, planning and the built environment', *Environment and Planning C: Government and Policy* 11, 2: 213–31.

—— (1993b) 'Creating barrier-free environments', *Town and Country Planning* 61, 10: 278–81.

Jary, D. and Jary, J. (1991) *Dictionary of Sociology*, London: HarperCollins.

Jencks, C. (1973) *Le Corbusier and the Tragic View of Architecture*, London: Allen Lane.

Jones, B. (1990) *Sleepers, Wake! Technology and the Future of Work*, Melbourne: Oxford University Press.

Klugman, K., Grant, B., McGuigan, A. and Lamberton, D. (1991) *From Isolation to Opportunity: Applying Communication and Information Technology to Facilitate the Placement of People with Physical Disabilities into Open Employment*, Working Paper 1991/9, Melbourne: Centre for International Research on Communications and Information Technology.

Kraut, R. E. (1989) 'Telecommuting: the trade-offs of home work', *Journal of Communication* 39, 3: 19–47.

Kridler, C. and Stewart, R. K. (1992a) 'Access for the disabled 1', *Progressive Architecture* 73, 7: 41–2.

—— (1992b) 'Access for the disabled 2', *Progressive Architecture* 73, 8: 35–6.

—— (1992c) 'Access for the disabled 3', *Progressive Architecture* 73, 9: 45–6.

Le Corbusier, Charles-Edouard Jeanneret (1946) *Towards a New Architecture*, London: The Architectural Press.

—— (1971) *The City of Tomorrow and its Planning*, London: The Architectural Press.

Mandel, E. (1968) *Marxist Economic Theory*, London: Merlin.

Marx, K. (1981) *Capital: A Critique of Political Economy – Volume Three*, London: Penguin.

Murdock, G. and Golding, P. (1989) 'Information poverty and political inequality: citizenship in the age of privatized communications', *Journal of Communications* 39, 3: 180–95.

Mumford, L. (1966) *The City in History*, Harmondsworth: Penguin.

Napolitano, S. (1996) 'Mobility impairment', in G. Hales (ed.) *Beyond Disability: Towards and Enabling Society*, London: Sage.

Oliver, M. (1990) *The Politics of Disablement*, London: Macmillan.

Rifkin, J. (1995) *The End of Work*, New York: Putnam.

Roulstone, A. (1993) 'Access to new technology in the employment of disabled people', in J. Swain, V. Finkelstein, S. French and M. Oliver (eds) *Disabling Barriers – Enabling Environments*, London: Sage.

Scherer, M. (1993) *Living in the State of Stuck: How Technology Impacts the Lives of Disabled People with Disabilities*, Cambridge, MA: Brookline.

Scott, A. and Soja, E. (1986) 'Editorial', *Environment and Planning D: Society and Space* 4: 249–54.

Smith, N. (1984) *Uneven Development*, Oxford: Blackwell.

Soja, E. (1989) *Postmodern Geographies: The Reassertion of Space in Social Theory*, London: Verso.

Sokolowska, M., Ostrowska, A. and Titkow, A. (1981) 'Creation and removal of disability as a social category: the case of Poland', in G. L. Albrecht (ed.) *Cross National Rehabilitation Policies: A Sociological Perspective*, Sage: London.

Stewart, C. (1952) *A Prospect of Cities*, London: Longmans, Green and Co.

Toffler, A. (1971) *Future Shock*, London: Pan Books.

Topliss, E. and Gould, B. (1981) *A Charter for the Disabled*, Oxford: Basil Blackwell and Martin Roberston.

Vujakovic, P. and Matthews, M. H. (1994) 'Contorted, folded, torn: environmental values, cartographic representation and the politics of disability', *Disability and Society* 9, 3: 359–74.

Williams, R. (1980) *Problems in Materialism and Culture – Selected Essays*, London: Verso.

Young, I. M. (1990) *Justice and the Politics of Difference*, Princeton, NJ: Princeton University Press.

6

BODY TROUBLES: WOMEN, THE WORKPLACE AND NEGOTIATIONS OF A DISABLED IDENTITY

Isabel Dyck

Introduction

Throughout the social sciences the influence of cultural studies, feminisms and the challenges of post-modernism and post-structuralism have made space for sustained debate over the nature of human subjectivity, its constitution and its transformations. Difference, identity and the notion of the embodied self are being explored from various disciplinary perspectives, with geography focusing investigation on issues of space and place. The body, too, is attracting attention as the linkages between identity and the experience of specific spaces and places are theorised. Centring the body in inquiry in geography has primarily been through work of feminist geographers interested in the connection between the body and situated knowledges, and geographers concerned with questions of sexuality, but recently the 'deviant' body of disability has also emerged as a focus of investigation (see, for example, Dorn and Laws 1994; Moss and Dyck 1996; Park *et al.* 1998; *Environment and Planning D: Society and Space* 1997). This work points to the discursive construction of ideas about the body and its abilities, and how dominant representations may be negotiated and contested in the context of particular spaces and places as 'disabled' women construct the meanings and materialities of their everyday geographies.

The purpose of this chapter is to investigate such contestation as this occurs in the workplace, as the subjectivity of women with chronic illness, specifically multiple sclerosis (MS), is transformed as they struggle with their 'body troubles'. The women were living with a sometimes failing, often unreliable body and one difficult to control. The severity of the women's impairments varied, but at some point most had experienced a period of 'invisible' disability. That is, although women experienced symptoms that caused feelings of illness and prevented, or made difficult, certain activities, the women appeared healthy. While workplace experiences varied, identity issues were a common and sometimes a profoundly disturbing concern. The analysis is informed by a feminist materialism and work in geography concerned with the mapping of embodied subjectivities. I am particularly interested in how discursive constructions of the body, and particularly that of biomedicine, interplay with the environment – understood as a dynamic layering and interweaving of social relations and space (cf. Moss and Dyck 1996) – as women interpret and live their changing abilities. I have been concerned to ground the analysis in the everyday routines and spaces of the everyday lives of the participating women, and to draw on their voices in explicating my argument. Their experiences therefore are foregrounded, as described by them through in-depth interviews.

In the first section of this chapter I delineate the main theoretical ideas guiding the analysis. I then use data from the interviews to explore the women's changing subjectivity as this is grounded in the social practices and micropolitics of the workplace. The issue of identity and its negotiation is discussed with specific reference to concerns about disclosure of medical diagnosis, and the strategies used by women in negotiating the designation of 'disabled'. The study shows a destabilisation of women's identities as both representations of the body and the women's experiences of its corporeality shift, and the women's positioning in the labour force is open to reconstruction and redefinition. I argue that the paid workplace, as a collective institution and a collection of people, with its cultural codings as a space for the 'able-bodied', is a significant site in the negotiation of women's identity as 'disabled'. In the analysis I aim to show that such negotiation is grounded in the social practices of workplaces as 'places of risk' which themselves set parameters for the women's identity negotiation. Furthermore, these practices are embedded in social, political and economic relations that position 'disabled' people in a particular moment and status in the labour market in an exclusionary way. In the discussion, however, I suggest that rather than understanding the workplace as a static space with fixed boundaries, it is one whose meanings may be destabilised and contested in processes of the women's negotiation of their identity. I aim to show how the close and complex interweaving of the material body and its representations in biomedical discourse shape how the body may be simultaneously a site of oppression and a site of resistance as women reinterpret their 'place in the world'.

Bodies, identities, spaces: the embodied subject

In this chapter I am interested in how ideas about the body, identity and space nexus can inform the interpretation of the stories of women with multiple sclerosis and their experiences in the workplace. The body is receiving growing attention in discussions of social theory. Several different approaches have been used in theorising the body, ranging from essentialist understandings, through social constructionism, to the 'body as text' of post-structuralism. Bodies have been variously interpreted as physical capital or body as commodity, as loci of experience, as sites of discipline, as medicalised, and as bodies-without-organs (non-anatomical political surfaces of inscription) as the links between the corpo-reality of the body and its representations are explored (see, for example, Featherstone *et al.* 1991; Fox 1993; Frank 1989; Freund and McGuire 1991; Shilling 1993). As debate has engaged with different 'bodies', the corporeal body, its representations in discourse, and human subjectivity are shown to be complexly linked within a web of social processes, operating at different levels, and recursively engaged with spaces. The body as a static or essential 'object' is rejected, for it is understood as constantly in the making, embodying and contributing to social relations, and with its capacities constituted within cultural and historical specific moments (Grosz 1994; Shilling 1993). Shilling (1993: 4), for example, suggests that bodies are malleable and that the body is an ongoing project, never finished, but always 'in the process of becoming'.

Anti-essentialist feminist scholars have been interested in the ways dominant discourses, constructed within gendered power relations are part of this process of 'becoming', mediating women's experiences of the body and providing ways of interpreting such experiences. In addition to cultural discourses about femininity and masculinity, the body as 'text' or surface of inscription may also be 'marked' through racialising discourses, and those of class and, of interest in the context of this chapter, medicine which also marks the body through its material practices. Walkerdine (1995) comments on the multiple and historically specific scripts and stories circulating that may mediate experience and are constitutive of human subjectivity, so suggesting a multistrandedness of identities with a potential for fragmentation. Some strands will have greater salience and be drawn upon in different ways according to context, but uneven distributions of power will circumscribe the relationship between possible scripts and their embodiment through performative acts.

It is the linking of performative acts of particular scripts (Butler 1990) with the materiality of the everyday that helps us to understand the embodiment of culturally shaped subjectivities and knowledges. It is here that geographers have an important contribution to make, for embodiment takes place in space. As Philo (1996: 38) comments, embodied knowledges 'arise through our interactions with the environment – and which do so in distinctive ways if our bodies are differently "sexed" or if they are physically or mentally "sick" or "disabled" – to shape our senses of ourselves in time, space, period and place'. Pile and Thrift (1995: 11) describe the body as the 'spatial home' of the subject and subjectivity, moving in time and space, although not always freely. Limits on such movement are interpreted in terms of the intertwining of the material body and its representations in discourse, with Rose (1993: 32) further conceptualising the body within such processes as 'maps of the relation between power and identity'. Empirical studies are investigating the nuances and complexities of the interweaving of geographies and identities, showing the ways in which everyday spaces are variously implicated in the constitution and embodiment of human subjectivity. For example, various studies show the close intertwining of subjectivity, space and the body as spaces are experienced in different ways according to subject positionings such as gender, class, 'race', sexuality, age, and more recently disability (see, for example, Bell 1995; Dorn 1998; Dyck 1995; Katz and Monk 1993; Longhurst 1995; Moss 1997; Rose 1993; Valentine 1993). While geographers have emphasised and as Chouinard (1997: 379) reminds us, 'discursive and cultural codings of space are used to discipline, marginalise and exclude people with particular markers of differences', these studies also suggest that, just as identities are malleable and open to fragmentation, so meanings of space are potentially unstable and open to contestation. They question the usefulness of dichotomous categories, whether of persons or spaces, challenging us to think of 'third space' and the ongoing formations and renegotiations of boundaries of places and identities. Such processes suggest a world always in the making, but with cultural moments with a potential to 'fix' dominant meanings.

Such microscale studies, focusing on the materiality of everyday life, provide a useful entry point to the interweaving of the discursive and the material in investigating the formation of identities (Moss and Dyck 1996). They also permit exploration of 'competing' discourses as subjectivities are constituted and transformed. In this chapter, I am particularly interested in the tensions between the inscriptive processes of biomedicine as a powerful, cultural construction depicting the body as an 'object of science' (Fox 1993; Good 1994), other inscriptions of the body, and women's own experiences of living with chronic illness, as these interplay in reconstituting the body and subjectivity. Wendell (1996: 117) writes of the social and cognitive authority of Western scientific medicine in describing 'our bodies to ourselves and others' but its lack of ways of talking about and explaining the lived experience of illness or disability. As she wryly comments (1996: 122) of her own illness experience, 'my subjective descriptions of my bodily experience need the confirmation of medical descriptions to be accepted as accurate and truthful'. This comment picks up a central issue faced by the women in the study discussed here. Their bodies have been 'marked' through the language and practices of biomedicine, but this inscription interweaves with their bodily and social experiences following diagnosis in complex ways. While the analysis is grounded in the women's accounts, the following questions were posed: for women with chronic illness, how does bodily change threaten a continuity in self and social identification? how do such women live and renegotiate their subjectivity as corporeal changes are accompanied by the body's changing

122

meanings and representation in the workplace? if the body is in a state of ongoing transformation and definition, how do events challenging its continuity (as an embodied self) enter and change the routinely experienced links between materiality, representations and subjectivity in the women's lives? As described in the remainder of the chapter, the women's bodies were in a 'process of becoming' threatening former physical capabilities and self and social identities, and including a potential categorisation of 'disabled'. Yet it would be mistaken to view biomedical authority as uncontestable. There are possibilities for resistance. As women resisted and negotiated the formation of a disabled identity and the meanings attached to this, there was a recursive interplay between the material body, its discursive constructions, and the workplace as a specific site of interweaving social relations and space.

The study[1]

Qualitative methods were used to investigate women's everyday experiences of domestic and wage labour following diagnosis with MS. In-depth interviewing was chosen due to the method's ability to reveal the women's relationship to the complex layering of environment, through their accounts of their illness experience. Interviews were semi-structured in that various topic areas were to be covered, but the interviewers were guided by the issues raised by the women. The women were recruited from two sources: a local branch of the Multiple Sclerosis Society and a neurological clinic specialising in MS.[2] The analysis here concerns thirty-one women in either part-time (twelve) or full-time (nineteen) employment. They ranged in age from 25 to 49 years. Diagnosis of MS is predominantly associated with white people; this biological 'fact' was reflected in that the women recruited were white. Fourteen were married or lived in a stable heterosexual relationship, seven were divorced or separated, and ten were single. The women's employment ranged from service, sales and clerical work to professional occupational categories, although most of those still in employment worked in managerial, technical or professional occupations. All lived within Greater Vancouver, British Columbia, Canada. Interviewing produced many pages of detailed accounts of the women's day-to-day experiences in their home, neighbourhood and work environments following diagnosis. A thematic, interpretive analysis was employed, involving careful reading of all transcripts and a constant comparison across the women's accounts by the author and a co-investigator interested in policy issues. In this chapter I focus on the issue of disclosure and on women's restructuring of the workplace as they struggled with their bodily changes.

Limited space precludes an extensive drawing on verbatim quotes from the interview transcripts; those I use are chosen as typical of women's concerns, although the particularities of the women's situations differed. As in any research concerned to link the particular with broader political economy relations, an attempt is made to balance the commonalities in women's experiences and the specificity of context in the analysis. However, in the interests of maintaining the women's anonymity details of context and demographic characteristics are not provided for each individual. Consistent with the aim of drawing on the women's voices in explicating my argument, however, I introduce the main issue addressed in the chapter with interview excerpts then contextualised in a brief summary of the two women's current situation:

Helen: MS is very inconvenient. It's inconvenient because it doesn't show.

Interviewer: Okay, what do you mean then?

Helen: Well, I mean you can sit there, I mean I look perfectly healthy.

Interviewer: Right.

Helen: You don't know that my right leg's numb and maybe half my face is numb, that if I close my eyes I'll fall over. You don't know that I am tired, it doesn't show.

Elaine: I don't have a problem with it right now . . . I only have a slight problem on my left side and nobody can notice it. And my speech is slurred occasionally but it's usually just when I'm tired.

Multiple sclerosis is a chronic and often progressive neurological disease which may be manifested in a variety of symptoms, but commonly involves profound fatigue and sensory and motor disturbances. Both 'Helen' and 'Elaine' were employed full time in positions in which career success and security were tied to high performance standards in the workplace. Helen was married with two teenagers, while Elaine was unmarried and living alone. The women talked about themselves as 'invisibles', a category used by women in the study to distinguish between those with hidden disabilities and those who had clearly observable manifestations of the disease. Both recounted their struggles in maintaining their performance and position in the workplace. These struggles were not confined to physical difficulties in carrying out job-related tasks but extended to identity issues. While Helen's disability was in doubt, as her colleagues found it hard to understand she was sick when her body looked 'perfectly healthy', Elaine in contrast was more concerned with appearing well and concealing disabling symptoms. She had not disclosed her illness in the workplace. Like other women in the study, their present concerns are subsequent to a process of biomedical inscription, starting with diagnosis, in which their body was defined as 'diseased' with a considerable, but uncertain, potential to become disabled.[3]

Rescripting the body: changing possibilities, changing subjectivities

A recurrent theme in the women's accounts of their workplace experiences and decision-making was that of the medical uncertainty accompanying diagnosis. Diagnosis represented a point at which a biomedical script was drawn on in depicting and explaining a woman's changing experience of her body. For most of the women considerable medical uncertainty had accompanied such naming and explanation of their symptoms and ongoing illness experience. Diagnosis had often followed several months or even years of puzzling, sometimes transient, and debilitating symptoms. This period of illness was described by the women as a time when they felt a loss of control of their body. Eventual diagnosis brought relief to many women, in the sense of legitimising women's own illness experience and one that might have been doubted by others, whether physicians, family members, friends, or work colleagues, but the biomedical script also brought continuing uncertainty. Although it is known in biomedical science that MS is a progressive neurological disease with the potential for severe disability, there are various courses the disease may take. Women may have periods of remission with minimal or no symptoms or their bodily capacities may deteriorate steadily although with an unknown temporality. There is no known cure, although management strategies may relieve symptoms.

The women's 'reading' of the biomedical discourse describing their bodies and the material changes in them, however, varied and was interpreted within the context of their everyday relationships and day-to-day routines. Some talked explicitly about the tension between the text of MS and their own experience: while symptoms might match the information conveyed in the medical explanation, their meaning to women was not adequately covered by a medical focus and its terminology. The inadequate fit between the authoritative script of biomedicine and women's own bodily experiences was an issue that became a particular struggle in the workplace. The meaning of hidden disability varied for the women, but for all there was a destabilisation of identity that came into sharp definition in the workplace. The next sections focus on how women negotiated changes to the body's corporeality and the meanings conveyed by a biomedical scripting as their positioning in the labour force and its associated rewards were threatened.

The workplace as a place of risk: fragmenting identities, destabilised meanings

Various strategies were employed by the women as they negotiated the changes in their material body and its associated meanings in the home, in large part away from the public gaze (Dyck 1998). In contrast, the workplace, reflecting the organisation of commodified labour within the social relations of capitalism, was for many women one where performative capacities were at a premium, and often visible to others. Abilities making up workplace performance were written into workplace social practices, whether informally or through policy. For instance, the lack of seating for retail store sales clerks or supermarket cashiers means that women having difficulty standing for long periods will lose such employment. In other work environments informal social practices that have become normative cues for behaviour, such as climbing stairs to the office instead of using the elevator, or a high-paced work 'culture', pose a threat to women's identity performance in the workplace. Struggles around the meaning of having MS in the workplace, as this related to work abilities, was a point of tension for many women, exemplified in one woman's comment:

> It's up to me to say whether I'm being affected to the point that I cannot do my job, not for them to say that. So, I just don't feel that I should be sick at this point in my life. Because I'm not ready, you know?

Although the demands of specific settings varied widely, for most women the workplace became a place of risk. Not being able to perform 'as usual' potentially threatened women's financial stability, and consequential access to a range of resources and opportunities, including housing. A woman's position as a social being in the world was also challenged. Threats to self-identity were commonly voiced. Women were concerned about being treated differently by others, and struggled with their own identity as an 'able self' as their performative acts no longer consistently matched their former interpretation of their 'place in the world'. Such an interweaving of concerns was expressed by one woman in the following way:

> I enjoyed me immensely and . . . you know, one of my big tasks since I've been diagnosed has been trying to . . . deal with changing that image . . . I just don't want to identify with the disabled. But I think that that's only part of it . . . I think that there's discrimination out there that I don't know what to do with.

Other women drew a close association between their ability to work and both their self and social identity. One woman who had returned to employment stated:

> Working part time gives you this whole – not only a little bit more money, but it gives you this whole thing . . . because we have this culture that if you don't have a job . . . you're not a person. And then if you tell that you don't have a job because you're on long-term disability – I mean you're even less of a person . . . Even if you're working 15 hours . . . you have a place – you have an employer and you have a job and you have this thing that you do.

126

The intertwining of an able identity and work participation is implicated in the resources available to women. The material and financial rewards of a place of employment are jeopardised in part by physical limitations that circumscribe which job tasks women can or cannot continue to do. In addition, the representation in discourse of the body as diseased comes into play in the meanings attributed to women's performance in the workplace. The biomedical 'scripting' of the body provides an authoritative set of descriptions and meanings through which to interpret the women's struggles with their bodies. Women with hidden disabilities were often unsure of the implications of declaring their biomedical diagnosis, expressed in a common anxiety and dilemma surrounding the issue of disclosure.

Disclosure of diagnosis represented a pivotal and dynamic moment in which a 'marked' identity, that of disabled, had a potential to be 'fixed' with uncertain consequences for the women. While some women had greater control over their work environment than others, commonly cited fears associated with disclosure in employment situations included being unable to get work, losing a job, failing to gain promotion, or jeopardising eligibility for disability insurance or pensions. Yet women were also aware that a diagnosis may be drawn on in different ways as they negotiated their position in the labour force. It did not necessarily convey a single meaning. Its public knowledge may constitute a threat to continued employment, with consequent social marginalisation, but may also provide women with access to a social safety net of disability benefits and pensions, and sometimes access to help on the job allowing the completion of work tasks. In the next sections I explore women's management of meanings about their destabilised identities and 'diseased' bodies. Central to the strategies of most was resistance to the dominant biomedical conceptions of their bodies as diseased and particular interpretations of this in the workplace.

Negotiating the workplace and deferring meanings

Industrialisation, capitalism and the commodification of labour have been important in shaping the conditions under which dominant meanings of health, illness and disability and their categorisation are constructed (Barton 1996; Oliver 1990; Park et al. 1998; Zola 1991). Post-structuralist writing, however, suggests that meanings can be deferred and transformed, with a potential for their re-territorialisation as new meanings to be acted upon and bodies and spaces to be reinscribed (see, for example, Fox 1993, for the case of medicine). The women's accounts of their workplace experiences indicate slippage of meanings around the dichotomous categories of able/disabled, over which they struggled and in part managed through spatial strategies.

Helen, quoted at the beginning of this chapter, found her work colleagues underestimated or doubted the severity of her illness due to the invisibility of her symptoms, despite their legitimation through a medical diagnosis, but other women were concerned to remain invisible and hide their diagnosis. Non-disclosure and the concealment or management of symptoms were common strategies employed by women to manage this 'secret knowledge' and defer its meanings, as they negotiated both their ability to work and the threat to their able identity. Women may have disclosed their diagnosis in other contexts, for example to family members and friends, but disclosure in the workplace was resisted by women attempting to maintain the integrity of their existing social identity. This was particularly the case when there was a disjuncture between what women felt they were able to do and the meanings evoked by a diagnosis that suggested disability. Women whose symptoms were in remission might work for long periods with no or minimal symptoms, but exacerbations made concealment more problematic or impossible. Visibility of symptoms was situational too, occurring for some women only with tasks requiring mobility, or following the accumulation of physically demanding tasks over a day. The importance of appearing able was voiced by several women. One, for instance, talked of being upset when she had gone through periods of having to use a cane, 'I always look the same, but of course I have a cane and I'm limping and I don't look like I know what I'm doing, right?'

The ability of women to employ concealment strategies was linked not only to the severity of their symptoms, but also to the temporal-spatial organisation of work tasks. The management of space was integral to concealing deteriorating physical capacity. This included strategies such as avoiding walking and climbing stairs when possible, and the organisation of work tasks so that they were spread over more than one workspace, sometimes including the home. Each strategy involved the management of the body in the workplace in ways that reduced attention to its limitations. These are illustrated through examples from several women's descriptions of dealing with the issue of disclosure. Non-disclosure at work for one woman, for example, was facilitated by work assignments that took place in a variety of different spaces, as well as flexible work hours. Her work involved meeting clients in their workplaces, and her control over work scheduling allowed her to pace her travel and appointments in such a way that she was able to appear well and work competently. Her main problem of fatigue was accommodated by this scheduling and use of space. Another woman who had difficulty with handwriting took some of her work home, where she transcribed shakily written notes she would previously have given to a secretary. Women in various forms of higher level sales involving work with professional clients used a variety of spatial strategies in avoiding the appearance of being disabled. One woman always allowed clients to leave the office ahead of her to cover the difficulty she had standing up from her chair and her limping. Another avoided lengthy tours of work sites when feeling fatigue by claiming time constraints and the need to get to another appointment.

Attempts by women to maintain the appearance of an able identity through substituting one way of moving through the physical environment of the workplace by another might be jeopardised, such as when the alternative was not consistent with workplace norms. For example, using an elevator instead of stairs was a common way of managing the physical environment for women with fatigue or mobility difficulties, but often women felt a need to provide a rationale. Similarly, being able to 'fake it' as one woman put it, sometimes involved a withdrawal from workplace social activities that were considered part of workplace collegiality. One strategy adopted in providing a rationale for lack of participation in an activity was claiming another illness or physical problem as the cause of difficulties in performance, which was perceived by women as less stigmatising than MS. A woman, for example, whose fatigue problems could be handled through a routine management of time and space, had difficulty when she was away from home at conferences. She explained her need to go to bed early or lie down and rest during the day as due to a back problem, a reason she saw as more socially acceptable than MS. Another, whose increasingly unsteady walking became a problem in a work environment that involved a good part of the day walking, standing and the use of stairs, avoided disclosure of her MS for several weeks as a broken ankle precluded these activities. Flu was the legitimising reason given by another woman for time taken off work due to an exacerbation of MS symptoms. Part of her job required walking outside from building to building to transfer information and a co-worker took over this task when she was first back to work, on the basis of her still recovering from flu. Another who had been asked what was wrong with her legs, replied 'I twisted my ankle'.[4]

The potential stigma of MS, together with uncertainty concerning access to long-term disability insurance for some women, were considerations for women struggling with the issue of disclosure. Yet, unless women enjoy a long remission of symptoms or symptoms are minimal, it is unlikely disclosure can be avoided in the longer term, particularly when women have little control of their work conditions. Furthermore, women in jobs requiring long periods of standing or other physically demanding activity had few options to use space in a way conducive to minimise symptoms or their visibility. By the time of the interviews most of the women had, in fact, disclosed and their diagnosis of MS was known in the workplace.

Negotiation and contestation of workplace meanings

Following disclosure some women who appeared healthy found their claims of illness doubted by unsympathetic co-workers or supervisors, but more usually women remaining in employment had found supportive work colleagues and respect for their self-declared limitations. This is to be expected in the context of this study in that those able to retain their job were necessarily working under conditions where such support allowed this. Disclosure, however, was often selective with only an employer, supervisor or a few close colleagues being privy to this knowledge. This was particularly the case when women were in an environment with many co-workers or where they were able to carry out job tasks without major adjustments to tasks or the physical arrangement of the environment.

The demands of some jobs, particularly in sales and some service occupations, precluded continued employment for some women who had little option of occupational change without further training. For example, a hairdresser was forced to quit her job whereas a nurse with similar physical limitations was able to transfer to a job with lighter physical requirements. Being able to work part time or with flexible hours allowed some women to remain in the labour force. For others spatial strategies continued to be an important part of being able to retain employment in a variety of occupations, although now with the knowledge and cooperation of employers or colleagues. One woman, for example, who had difficulty walking omitted coffee breaks, taking instead a long lunch break and so decreasing the amount of walking required to reach the coffee room. Another working in a secretarial position had been able to continue to work through a rearrangement of the photocopying tasks required, accumulating these and doing them once a day and so reducing the amount of walking she had to do. Some workplaces had a room or a couch available for resting, but few made use of this provision. As one woman said, 'I have to give 110 per cent just to prove that I'm really with it and I can really do it: I wouldn't feel comfortable [lying down].' Her comment echoed other women's continuing desire to be seen as able to do their jobs well, although accommodations might have been made.

The sociopolitical organisation of the workplace also became important, with distinctions between unionised and non-unionised workplaces, public corporations and private sector companies, and with public service institutions concerned with health and education forming a further dimension in shaping women's experiences. Women most vulnerable to loss of employment were those in non-unionised, private sector jobs, of low seniority, and with little control in the scheduling or organisation of job tasks. One of the women, working in retail sales, for example, talked of the store management's gradual reduction in her hours and her feeling that she had been harassed out of her job. Other women with greater control of their work environment were more able to contest and negotiate normative meanings and practices of the workplace and, through this, their own position in the labour force. For example, an elementary school teacher discontinued supervising extra-curricular activities and resisted the scheduling of informal or formal meetings with staff, students or parents in lunch times or coffee breaks, which she needed to preserve as rest times. Those women most able to restructure their work environments in order to maintain their relationship to the labour force and its material and social rewards were those in employment situations willing and able to provide flexibility, whether in hours, pace or the organisation of workplace tasks. Seniority and 'track record' on the job were also important in some workplace settings in influencing responses to women's changing abilities. One woman, for example, who had worked in community health services with most of her work involving travel to different community settings had a job created for her which drew on her experience and skills, but was office based.

Workplace organisation, located within social relations shaped by political economy, the social practices of the workplace, and women's own work histories all circumscribe the range of options they have in responding to their illness experience. Furthermore, women are positioned differently in the relations and distributions of power that shape their future employment opportunities. Those with higher levels of education and jobs in professional and managerial occupations generally were more able to maintain their employment, at least for a period. Class positioning was therefore a dimension of women being able to control conditions in the workplace and being less vulnerable to surveillance practices. However, the situation is more complicated than one of class for some occupations with high performance expectations which require visible performance as an 'able' worker, just as in lower paid service work.

The workplace, too, may be open to reinterpretation, although again this is more possible for women with greater control over work conditions. For many of the women in this study, for example, negotiations of tasks, hours and space were ways of contesting dominant meanings of the workplace as a place for the 'healthy', able body as women redefined their relationship to their work environment. At this level women acted as individuals. One woman's situation, however, demonstrates the collective politicisation of her struggle in the workplace. While unusual among the women in the study, her situation helps to show the grounding of processes involved in the formation of a disabled identity and its embodiment in the contextual specificity of everyday material practices. It also demonstrates that dominant workplace meanings and practices can be contested, resulting in a redefinition of the workplace and the relationship of those with a 'disabled' identity within it. Her account includes, and brings together, components of other women's stories as it spans the time from her initial employment, her struggles with her body and identity, and her reinscription of herself and the workplace. As such, it acts as an empirical summary of the main issues of the paper. It speaks to commonalities among the women, while acknowledging the particular linkages of their personal histories with the work environment as a physical and social space located within wider political economy relations.

The woman, a teacher in a higher education institution, had been diagnosed with MS before she applied for her job. She did not disclose her diagnosis, although she felt uncomfortable not doing so at the time. She had weighed the pragmatism of getting a job, believing that knowledge of her disease might preclude this, against her personal scruples. She believed herself capable of doing the job and managed this through restricting her social life, cutting out extra-curricular college events and using concealment strategies in the workplace. Her symptoms worsened, however, and she could no longer work in what was a full-time position. She commented on her experience of working prior to this time:

> I'd been there three years with nobody knowing and that felt awful . . . withholding, walking around with that knowledge, and worry . . . I made up little stories about – why you don't take the stairs, why you don't go to the dance, why you don't do this and that . . . the answer to the questions always in my mind was MS, but that's not something I shared . . . so the whole thing felt quite foreign to me and to what I knew about myself, to what I'd done before.

Following a later improvement in health and a reorganisation of living space that reduced the demands of her household labour she felt able to work again, but part time. However, the only position to become available was defined as full time. With the encouragement of colleagues in the same unionised workforce she applied for the position but was not offered it. She said, 'I was the most senior person applying . . . [but] because I couldn't work full time they offered it to another person, so then we grieved it.' She and the union embarked on a grievance process, wanting to establish that the only reason for her not getting the job was that she could not work full time. After a long process involving the aid of a lawyer as several steps of the grievance process were gone through, eventually the grievance was resolved in her favour, and she received a contract for a permanent part-time position. She commented on the process, noting that she could not have won the case without the union but also saying:

> The only reason I didn't have a permanent job was not because of my ability or my experience or my recommendations, it was simply because I couldn't work full time . . . And that's what was so hard for me . . . because I had a career full of successes . . . and now here I was and I couldn't even get somebody to give me a permanent job . . . [I] felt so belittled by everything. At the same time the disease is ravaging my body, it's also ravaging my mind and my spirit, so I really needed to win that one. And I'm still feeling good about that . . . not marginalised, I'm legitimate.

This woman, like others in the study, had initially negotiated the work environment through her body in the form of concealment strategies, but once her ability to work full time became unrealistic for her, her body became a politicised site of struggle. The scripting of her body by biomedicine and disclosure receded as issues of importance become irrelevant over time, to be replaced by a struggle over the tension between her bodily capacity and the normative, performance demands of the workplace. The struggle for employment was a struggle over both identity and body. This case of the redesignation of an employment opportunity to accommodate a woman's inability to work full time serves as apt example of the contestation of socially imposed meanings of the workplace and resistance to key players who order and interpret workplace practices. The workplace was an important site in the transformation of the woman's subjectivity and its embodiment, her winning of the case interrupting the formation of a disabled identity. Furthermore, just as her 'disabled' identity was reinscribed as 'able', under the specific and negotiated conditions of employment, so too the meaning of the performative standards of the workplace were redefined.

Discussion and concluding remarks

It is likely the women of the study would agree, at least to some extent, with Wendell's (1996) comment about the need for subjective descriptions of bodily experience to be confirmed by medical descriptions in order to be accepted as valid and truthful. Certainly those that 'look perfectly well' need this legitimation to have their illness claims taken seriously, whether in negotiating workplace performance or gaining access to resources reserved for the disabled, such as long-term disability insurance or the Canada Pension Plan. For women with MS whose symptoms are in remission or transient, and who are able to work with the support of co-workers and various symptom management strategies, the potential or experienced stigmatising effects of their diagnosis is resisted by its concealment when possible. For them, disclosing their biomedical scripting may close off employment opportunities or open up possibilities for renegotiating their work tasks and conditions, depending on how employers 'read' and interpret such an inscription in relation to the work practices of a particular workplace. Disclosure of a diagnosis of MS was an important moment in the reconstitution of women's subjectivity. A disabled identity may become 'fixed' but with different consequences for women as they negotiate their position in the labour force.

Workplaces are recursively implicated in the reconstitution of subjectivity as women become defined as 'disabled'. As environments comprised of a layering of social relations and spatial organisation they provide the specificities of context for that cultural 'moment' of the discursive and material inscription of a disease category on women's bodies. However, the salience of a disabled identity may vary from setting to setting. Furthermore the concealment strategies women use to 'cover' potentially stigmatising symptoms and the case of the woman who worked with a union in resisting her social marginalisation indicate that the meanings and boundaries of what constitutes the workplace are potentially unstable. Hegemonic notions of the workplace and appropriate workers can be contested.

The women's accounts of their struggles with their body troubles support the argument that the corporeal body is continually in the process of 'becoming' as suggested by Shilling (1993), and always interpreted through available scripts, whether these be about ableness/disability, class, gender, 'race', sexuality, age, religion, caste or other axes of social differentiation accompanied by different insertions in distributions of power. These culturally produced scripts may be powerful, attaining a hegemony of understanding of the body, inclusive of ideas of ability or disability, but they may be resisted as women negotiate their identities and bodies in the materiality of their everyday lives. Yet there are limits to this resistance through individual body politics. Those inserted differently in distributions of power, such as supervisors, employers and disability insurance assessors, are in a position to use their interpretations, backed by a biomedical inscription, to contest or support women's attempts to renegotiate their position as participants in the labour force. Even when the body becomes a collectively politicised site of resistance, a woman will be constrained in the extent to which she can reinscribe herself in relation to a work environment. Relations of political economy forge the conditions of the workplace, although inclusive workplace policy and affirmative action may protect the position of workers in some settings. The divide between able to work or 'disabled' is one imposed through the machinations of social policy and private insurance schemes, rather than one that reflects the bodily experience of chronic illness as talked about by the women of this study.

To be able or disabled carries different connotations for the women. It is not a unitary experience, and how the biomedical script is drawn on varies, and conveys different meanings in different sites of interaction. As women struggle to defer meanings about their (dis)ability they do so within discursive and material practices, as played out in the particularities of time and space. The contingencies and local social and material practices of specific workplaces frame the fixing of the category of 'disabled', and often place a woman in the oppositional category she sought to avoid. Such practices producing subjectivity, and its embodiment as microscale geographies and identity intertwine, are located within cultural and historical specificities (Walkerdine 1995). 'Coming out' as disabled is not just a personal moment but one embedded in processes of categorisation that rely on dualistic epistemologies that then maintain categories (Pile 1994). As Natter and Jones (1997) comment, the category is a generalising and homogenising moment which serves to constitute both self and other. To be both able and disabled, or situationally or variably so, does not find a ready place within practices informed by biomedical representations of health, illness, disease, disability and body. As women make claims to the paid workplace through spatial practices, and experience exclusions from it, they are therefore also contesting claims of authenticity about the world. Their bodies have been marked through the language of biomedicine, but their performative abilities are reinterpreted as this inscription interweaves with their bodily and social experiences in complex ways in particular workplace contexts.

Acknowledgements

The research was funded by grants from the British Columbia Health Research Foundation and the Social Sciences and Humanities Research Council of Canada. The co-investigator of the study was Dr. Lyn Jongbloed, School of Rehabilitation Sciences, University of British Columbia. Both investigators carried out some interviews but the majority were conducted by the research assistant to the study, Roberta Bagshaw. Our greatest debt is to the women who generously gave of their time and energy in participating in the research.

Notes

1 A second, separate, phase of the study consisted of a questionnaire survey, which derived its questions from this qualitative study. This is reported separately in Jongbloed (1996).

2 Reflections on the recruitment process in the context of feminist methodology can be found in Dyck (1996).

3 I use the terms sick and ill(ness) to indicate the experiential dimensions of symptoms associated with MS. Disease refers to the diagnosis of MS as represented in biomedicine. Disability is used to indicate limitations in performance in the context of the specificity of a workplace.

4 The women usually had neither the time or the energy to be involved in support groups. They were commonly isolated from other women with MS with whom such strategies might have been shared.

References

Barton, L. (ed.) (1996) *Disability and Society: Emerging Issues and Insights*, London: Longman.

Bell, D. (1995) 'Pleasure and danger: the paradoxical spaces of sexual citizenship', *Political Geography* 14: 139–54.

Butler, J. (1990) *Gender Trouble: Feminism and the Subversion of Identity*, New York: Routledge.

Chouinard, V. (1997) 'Making space for disabling difference: challenging ableist geographies', *Environment and Planning D: Society and Space* 15: 379–87.

Dorn, M. L. (1998) 'Beyond nomadism: the travel narratives of a "cripple"', in H. Nast and S. Pile (eds) *Places Through the Body*, New York: Routledge, 183–206.

Dorn, M. and Laws, G. (1994) 'Social theory, body politics, and medical geography: extending Kearns's invitation', *The Professional Geographer* 46: 106–10.

Dyck, I. (1995) 'Hidden geographies: the changing lifeworlds of women with disabilities', *Social Science and Medicine* 40: 307–20.

—— (1996) 'Whose body? Whose voice?', *Atlantis* 21: 54–62.

—— (1998) 'Women with disabilities and everyday geographies: home space and the contested body', in R. A. Kearns and W. M. Gesler (eds) *Putting Health into Place: Landscape, Identity and Wellbeing*, Syracuse: Syracuse University Press, 102–09.

Environment and Planning D: Society and Space (1997) 'Special issue: Geographies of Disability' 15: 379–480.

Featherstone, M., Hepworth, M. and Turner, B. (eds) (1991) *The Body: Social Process and Cultural Theory*, London: Sage.

Fox, N. J. (1993) *Postmodernism, Sociology and Health*, Buckingham: Open University Press.

Frank, A. W. (1989) 'Bringing bodies back in: a decade review', *Theory, Culture and Society* 7: 131–62.

Freund, P. and McGuire, M. (1991) *Health, Illness and the Social Body*, Englewood Cliffs, NJ: Prentice-Hall.

Good, B. J. (1994) *Medicine, Rationality and Experience*, Cambridge: Cambridge University Press.

Grosz, E. (1994) *Volatile Bodies: Toward a Corporeal Feminism*, Bloomington: Indiana University Press.

Jongbloed, L. (1996) 'Factors influencing employment status of women with multiple sclerosis', *Canadian Journal of Rehabilitation* 9: 213–22.

Katz, C. and Monk, J. (1993) *Full Circles: Geographies of Women over the Life Course*, New York: Routledge.

Longhurst, R. (1995) 'The body and geography', *Gender, Place and Culture* 2: 97–105.

Moss, P. (1997) 'Negotiating spaces in home environments: older women living with arthritis', *Social Science and Medicine* 45: 23–33.

Moss, P. and Dyck, I. (1996) 'Inquiry into environment and body: women, work and chronic illness', *Environment and Planning D: Society and Space* 14: 631–783.

Natter, W. and Jones, J. P. (1997) 'Identity, space and other uncertainties', in G. Benko and U. Strohmayer (eds) *Space and Social Theory: Interpreting Modernity and Postmodernity*, Oxford: Blackwell Publishers, 141–61.

Oliver, M. (1990) *The Politics of Disablement*, London: Macmillan Press.

Park, D. C., Radford, J. P. and Vickers, M. H. (1998) 'Disability studies in human geography', *Progress in Human Geography* 22, 2: 208–33.

Philo, C. (1996) 'Staying in? Invited comments on "Coming out: exposing social theory in medical geography"', *Health & Place* 2: 35–40.

Pile, S. (1994) 'Masculinism, the use of dualistic epistmeologies, and third spaces', *Antipode* 26: 255–77.

Pile, S. and Thrift, N. (eds) (1995) *Mapping the Subject*, London: Routledge.

Rose, G. (1993) *Feminism and Geography*, Minneapolis: University of Minnesota Press.

Shilling, C. (1993) *The Body and Social Theory*, London: Sage.

Valentine, G. (1993) '(Hetero)sexing space: lesbian perceptions and experiences of everyday spaces', *Environment and Planning D: Society and Space* 11: 395–413.

Walkerdine, V. (1995) 'Subject to change without notice: psychology, postmodernity and the popular', in S. Pile and N. Thrift (eds) *Mapping the Subject*, London: Routledge, 309–32.

Wendell, S. (1996) *The Rejected Body*, London: Routledge.

Zola, I. K. (1991) 'Bringing our bodies and ourselves back in: reflections on a past. present and future "medical sociology"', *Journal of Health and Social Behaviour* 32: 1–16.

7

WORKSPACES: REFIGURING THE DISABILITY–EMPLOYMENT DEBATE

Edward Hall

There is no doubt that disabled people are in a disadvantaged position in relation to employment in Britain. More disabled people are unemployed, in lower status occupations, on low earnings, or out of the labour market altogether, than non-disabled people (Martin *et al.* 1989; Berthoud *et al.* 1993; *Labour Force Survey* 1997). However, it is less clear why this situation has occurred, how it happens and what makes change so difficult. This chapter aims to provide a starting point for understanding this situation. There are three parts to this 'refiguring' of the disability–employment debate: first, a questioning of the two main interpretations of the disability–employment relation; second, introducing the social theorising of 'the body' into both the disability and the employment debates; and third, through a case study, using the theory of the 'body' to develop an embodied and spatialised understanding of the disability–employment debate.

There is an important political context to this discussion. The present Labour government's 'Welfare to Work' policy includes an initiative to get more disabled people into employment, the aim being to, 'reduce welfare dependency, encourage the work habit and self-reliance and reduce the massive social security bill' (*The Guardian* 1997). At the same time, the recent *Disability Discrimination Act* (1995) has begun to take effect. The central section of the Act makes discrimination against disabled people in relation to employment illegal for the first time. Although there has been criticism of its lack of 'teeth' and the low number of successful cases (*Labour Research* 1997), the Act has put the issue of equal rights for disabled people on the policy agenda. Together these events make the issue of disability and employment not only important for the disabled people actually involved in work, but also raise more fundamental questions about the position and identity of disabled people in British society.

Understanding disability and employment

Although it is impossible to get a clear picture of the overall position of disabled people in employment – for reasons of differences in the definition of disability and the unevenness and poor quality of data collection – it is none the less useful and important to sketch a rough picture of the situation.

There is a consensus that of the two million disabled people of working age in Britain, only 700,000 or 30 per cent are in employment (Martin *et al.* 1989; Prescott-Clarke 1990; *Labour Force Survey* 1997). Further, disabled people are over-represented in lower skilled work and under-represented in higher status jobs. As a consequence they earn significantly less than able-bodied employees.[1] Why are disabled people in such a disadvantaged position in employment? There are two main explanations: first, that disabled people's problems of impairment combined with the attitudes of employers exclude disabled people from employment; second, that disabled people are excluded from employment because of spatial and institutional barriers in the workplace, which are the result of the oppression of disabled people in Western capitalist society. It is important to note here that the first explanation remains the dominant discourse in Britain. I will now go on to discuss these explanations.

The first explanation is focused very much on the individual disabled person and their direct relationship with the employer. In this understanding, it is the individual's problems of impairment that determines their level of employment capability and they are chiefly responsible for their exclusion from employment. The employer compounds this by either discriminating directly against the disabled person by refusing them a job for reasons of their disability, or indirectly, by organising work or selection procedures in such a way that they do not favour disabled people. Such discriminatory practices are widespread (Morrell 1990; Graham *et al.* 1990). The solution to the disadvantage of disabled people in employment through this explanation would be, on the one hand, for the disabled person to improve their skills and educational qualifications, or recognise that they cannot work and so rely on welfare benefits for their income and, on the other hand, for employers to be educated about the abilities and needs of disabled people in employment. Although there are many practical and sensible notions in this explanation and solution, such as the very real discrimination that exists amongst employers, it has a major flaw. By focusing on the individual disabled person and the individual employer, it excludes any discussion of the role of wider societal processes in producing these attitudes and experiences. The responsibility lies firmly with the individual disabled person and the discriminatory employer.

The second explanation attempts to identify the underlying processes which exclude disabled people from employment, that is, what lies behind the employer's discrimination and the perceived inability of the potential disabled employee. This explanation proposes that the workings and structures of the capitalist economy 'produce' disabled people as abject and spatially and socially excludes them. Gleeson, who names this approach as 'historical–geographical materialist', argues, 'The materialist view locates the origins of disablement in capitalist society at the unseen and dynamic structural level of socio-spatial trans-formation: a dialectic of social and spatial change which has devalued the capacities of impaired people' (1996: 392). Gleeson's analysis builds on that of Finkelstein, who, in a now classic paper (1981), argued that the development of capitalism in Britain gradually excluded disabled people from employment as a society based on agricultural and small communities was replaced by the factories and machinery of industrial production. To quote Ryan and Thomas:

> The speed of factory work, the enforced discipline, the time-keeping and the production norms – all these were a highly unfavourable change from the slower, more self-determined and flexible methods of work into which many handicapped [*sic*] people had been integrated [in the pre-industrial period].
>
> (1980: 101)

The argument of these authors, although possibly over-deterministic in that it makes such a direct connection between the operation of the economy and the position of disabled people (and perhaps over-romantic about a 'golden age' of disabled people in work), does 'highlight the importance of the mode of production in significantly influencing perceptions and experiences of disability' (Oliver 1990: 29). And it does note that 'disability' was created as a category to differentiate between those who could work and those who could not (and therefore were in need of welfare) (Stone 1985). This separation and drawing of the boundary between disabled people and non-disabled people in relation to employment was a powerful force in the oppression of disabled people in wider society. As paid work is arguably central to social life, the physical and social exclusion of disabled people has meant two things: first, that other areas of life are also exclusionary (such as the significant institutional discrimi-nation present in the education and transport systems: Barnes 1991) and, second, that employment will continue to be exclusionary as disabled people are 'designed out' of work (Imrie 1996). There is a clear spatiality to this exclu-sion, whether it be in the design of the workplace or the separation of, and therefore the travelling between, home and work. The solution that arises out of this explanation is to change the whole way that disability is understood, and the associated structures and barriers which exclude many disabled people from employment.

The second explanation offers us perhaps the most satisfying understanding of the disability–employment relation, as it recognises both the powerful under-lying forces in society that exclude people from employment (i.e., it is not simply the attitudes of employers or poor access, or even the inadequacies of the disabled person) and the fact that 'disability' is a socially constructed concept, not a naturally given experience. However, it too has a significant weakness: it is fundamentally oppositional in that it sees disabled people and employment as distinct and impossible to reconcile (in that 'disability' was created to exclude people from employment) and in many ways is as deterministic as the first explanation. It also does not allow one to get a full understanding of the complexities of disabled people's experiences of employment. A different approach is needed, one that maintains the fundamentals of the materialist approach, but recognises the experiences of disabled people in work.

Significantly, these two explanations for the disadvantaged position of disabled people in employment echo the two main understandings of disability (described in the introduction to this book). The first explanation can be thought of as an example of the *medical* or individual pathology model of disability. That is, the reason that the disabled person is out of employment is due to their individual inabilities and the direct discrimination of the employer, and the solution lies within these individuals. The second explanation, with its focus on institutional discrimination and the structures and processes of society excluding impaired people from employment and so disabling them, is an example of the approach of the *social model* of disability (Bury 1996). The medical model is now rejected by most disabled people and academics, with the understanding offered by the social model favoured as it removes the individual responsibility from disabled people and makes change possible. Liz Crow, a disabled feminist writer, tells of her enthusiasm for this change:

> My life had two phases: before the social model, and after it. Discovering this way of thinking about my experiences was the proverbial raft in stormy seas . . . This was the explanation I had sought for years. Suddenly what I had always known, deep down, was confirmed. It wasn't my body that was responsible for all my difficulties, it was external factors, the barriers constructed by the society in which I live. I was being dis-abled – my capabilities and opportunities were being restricted – by prejudice, discrimination, inaccessible environments and inadequate support. Even more importantly, if all the problems had been created by society, then surely society could un-create them. Revolutionary!
>
> (1996: 206)

The social model has become the focus for a powerful social and political campaign, most importantly challenging the idea of what is a 'normal' person in society and making a distinction between 'impairment' and 'disability'. In doing so, it disconnects the biological and the social which, for disabled people, so long the subjects of medical intervention, was a major breakthrough. However, as Crow explains, the social model has its own weaknesses:

> So how is it that, suddenly to me, for all its strengths and relevance, the social model doesn't seem so water-tight anymore? It is with trepidation that I criticise it. However, when personal experience no longer matches current explanations, then it is time to question afresh.
>
> (1996: 207)

It is Crow's comment about 'personal experience' which is most telling. It is the exclusion of impairment – a central part of the social model approach – that Crow finds so problematic. French agrees, recounting her own experiences as a visually impaired person, 'I believe that some of the most profound problems experienced by people with certain impairments are difficult, if not impossible, to solve by social manipulation' (1993: 17). Crow argues that such experiences of disabled people must be recognised and impairment 'reclaimed' from the individual, medical understanding of disability (as 'determining' disability) and used to 'renew' the still highly relevant social model. The presence of impairment, of pain, of bodies and biology, is something that must be included in the disability debate. Morris puts this very effectively in her powerful and accessible book *Pride against Prejudice*:

> There is a tendency within the social model of disability to deny the experience of our bodies, insisting that our physical differences and restrictions are entirely socially created. While environmental and social attitudes are a crucial part of our experience – and do indeed disable us – to suggest that this is all there is to it is to deny the personal experience of physical and mental restrictions, of illness, of the fear of dying.
>
> (1991: 10)

By 'allowing' the experiences of people, of their bodies, of their weaknesses, their pain and pleasures, into the interpretation of disability, we can perhaps renew the social model of disability so that it reflects the everyday lives of disabled people.

The present ways of understanding disability and the disability–employment debate – as described above – suffer from the same oppositional and exclusionary problems. The following section is a proposal of how we could begin to rethink both the meaning of 'disability' and the disability–employment relation mindful of the above discussion.

It is important to state here, however, that both models and explanations do not have equal significance in the debate around disability and employment. The medical model of disability remains the dominant understanding of disability in British society and the direct discrimination explanation is the main way that employers and government understand the disability–employment debate. Yet this dominant discourse is not uncontested. Disabled people's organisations continue to campaign actively to secure equal rights and to redefine disability as a social construct.

There is a need to rethink both the concept of 'employment' and the concept of 'disability' in order to refigure the debate over the disability–employment relation. This rethinking has not only to find a satisfactory route that deals with the underlying socio-economic (capitalist) cause of disabled people's exclusion from employment, but also needs to incorporate the experiences and issues of being a disabled person in or out of work. The way I propose to do this is through the concept of 'the body'.

Disability, employment and the body

Geography has recently embraced social theories of the body, in particular in relation to gender (Rose 1993), sexuality (Bell and Valentine 1995) and identity (Pile and Thrift 1995). The latter argue that the body is a 'site of capture', a place where experiences and social and spatial processes accumulate and shape the present, future and even past experiences of being that 'embodied' person. Longhurst argues, '[All of the above authors are] playing a vital role in retheorising geography, that involves problematising the mind/body split and making the body explicit in the production of geographical knowledge' (1995: 102).

Within the study of disability, until recently, there has been little attention paid to the theories of the body, for good reason. For most of human history, people classified as disabled have been treated as if they were, literally, their bodies. This understanding was enhanced and deepened by the late nineteenth- and early twentieth-century expansion of the medical profession and the culture of disabilities as medically defined conditions of the body largely persists today. So any attention given to the body as an important part of a person's identity is greeted with, at best, scepticism by disabled people's groups.

The body has been understood in two main ways in academic debate and in wider society, which echo the 'individual/medical' and 'social' models of disability outlined earlier – the body as a 'biological' phenomenon and the body as a 'social' phenomenon (Shilling 1993). The biological view of the body as a 'natural' and pre-social entity understands that the capabilities and limitations of the body determine the state and development of society and social relations (Shilling 1993). This long-standing notion of biology determining behaviour has recently received a boost with the explosion of genetic theory (Rose 1997). A social constructionist understanding sees the body as being produced by the processes and relations of society. It is the understandings of the body that make it what it is: the body is the product of discursive practice (Foucault 1974). This understanding has had massive progressive implications for the social interpretation of issues such as gender, race and sexuality and the equal rights of women and black people. The direct causal connection made between the biology of a person's body and their social position has been challenged (Kaplan and Rogers 1990; Jordan 1982). However, this approach itself has been criticised, in a similar way to the biological understanding of the body, as reductionist, in that it has a focus exclusively on society (ignoring the influence of the biology of the body).

It is very difficult to think outside of this debate, or to put it differently, 'in-between' the two understandings of the body. There is a possible way, however, to map a path of the 'in-between', using the notion of 'embodiment' (Turner 1992). This understands the body as a biological or corporeal entity, but one that is involved with society. The body, in this sense, is neither determined by biology nor society, but absorbs and reacts to biological and social processes in a never-ending process. As Grosz puts it, 'far from being an inert, passive, non-cultural and ahistorical term, the body may be seen as a crucial term, the site of contestation in a series of economic, political, sexual and intellectual struggles' (1994: 19).

Disability and the body

We can see that by using theorists such as Turner and Grosz it is possible to respond to French and Morris's concerns about the social model effectively 'neglecting' the body. We can perhaps also begin to think about disability as an embodied experience, and understand that it is neither solely 'individual' nor 'social'. This is not to say that there is no such thing as 'disability', and that this approach somehow gets rid of or avoids the discrimination and difference that disabled people experience. Rather, analysing the 'body' can be a useful way of talking afresh about disability, outside the constraints of the two models. As Grosz argues, 'where one body takes on the function of the ideal [in this case the 'able' body] its domination may be undermined through a defiant affirmation of a multiplicity . . . of other kinds of bodies and subjectivities' (1994: 19). The notion of a disabled/able body dualism (where the latter is dominant and 'normal') can be disrupted by thinking of bodies as having lots of meanings, while recognising that the dualism has power and still operates. This 'embodied approach' can also be applied to employment, as the following section explains.

Employment and the body

It is perhaps obvious to say that employment is a bodily practice. Of course we use our bodies (here body/mind are understood in union) in our work. We walk, sit, carry, think, remember, organise and interact. Supposedly solely 'mental' jobs involve the physical body and purely 'manual' work necessarily includes the mind. *All* of the processes of work are inherently bodily, but then so are all aspects of life. We *are* our bodies and so whatever we do takes place through the body. It is perhaps this 'inherentness' that is the very reason why the body has received such little attention in the social sciences. When something is so fundamental and obvious it is often seen as 'natural' and outside of social scientific investigation. That the body is a social entity and is more than a natural, biological base for social action, is one of the major breakthroughs that cultural theorists and geographers have made in the last few years.

Amongst an increasing number of economic geographers there is a new willingness to consider the role of the body. McDowell, in her study of employees in merchant banks, makes the link between economic practices and bodily processes and this, she argues, allows one to make, 'a small and partial start to untangling the complexities arising from the fact that corporations are run by real people' (1993: 33). It is the employees and their bodies which 'make up' an organisation or company. McDowell's initial attempt to investigate, 'themes such as desire, subjectivity and the body, which may seem unfamiliar to economic geographers, are an important part of the understanding of the ways in which economic institutions operate' (1994: 729), and have become a significant area of economic geography research (see Lee and Wills 1997).

McDowell focuses on the growth of service employment and, in particular 'interactive service occupations', in which the 'the presentation of the self is part of the production and sale of the service' (1994: 732). Although the employee's body is central to *all* employment, these new occupations make the connection even closer because the employee becomes part of the product. Hochschild's now classic study of the employment practices of airline attendants (1983) shows how such employment not only involves the body, but also changes it. Her aptly titled *The Managed Heart* looked at how the attendants were trained to provide a service to airline passengers. The training assumed no 'natural' skill for service so their bodies had to be *worked on* to become the person for the role. This working on oneself was, crucially, not simply a surface process, such as a false smile or a few rehearsed welcoming words. Hochschild observed that both trainers and attendants rejected such 'surface acting', realising that to sustain such a type of service work they had to use 'deep acting' and actually 'become' the role. This control over the body, over the emotions, appearance, movement and overall 'deportment', becomes materialised in the body, from wrinkles on the face from frequent smiling, to maintaining a low body weight and bodily 'attractiveness', to an agile body to bend and stretch, move and stand, in the practice of the job. Mentally, the constant customer contact and maintaining a high level of 'happiness' leave impressions on the body, particularly through stress. The job is part of who the attendants are, it is 'embodied'.

Key here are the codes or rules of the body in employment: McDowell calls these codes 'bodily normalisation', which is 'often achieved by explicit regulations, codes and required behaviour, as well as by mechanisms of self-surveillance' (1994: 733). It is this combination of the apparent self-managed nature of employees in these new forms of employment and the set of regulations of employee performance that make up the complexity of much current employment. Within this context the experience and position of employees is represented, negotiated and materialised. McDowell's research shows how women were often understood as being 'out of place' in the male employment world of merchant banks, and their presence in these 'male' areas of work challenging to the dominant discourses/understandings. Disabled people are similarly seen as 'out of place' in most workplaces, their presence disrupting accepted notions of embodied employment. Space is clearly central to the organisation of employment. From the physical layout of workplaces, to the different spaces of work in a company, and the movement of employees between tasks, employment is thoroughly spatialised (Crang 1994).

The body is, I am arguing, the focus of the processes of employment as it is both the entity that makes things happen and, at the same time, the entity that is on the receiving end of such processes. The spatialised organisation of employment and the operation of the codes of behaviour and performance (discussed above) are central in the 'making up' of the employment experience (McDowell 1994; Crang 1994). The following section will use a case study of a site of employment to show how work space is involved in the refiguring of the disability–employment debate.

Work spaces of disability

'Central Bank'[2] is a major high-street banking company, with over 40,000 employees. Like all banks it has severely cut its staff and the number of its branches in recent years (*The Guardian* 1996). At the same time it has been restructured, giving greater employment and financial responsibilities to local managers, making high-street branches into 'shops' to sell its banking products and shifting processing and telephone enquiries to out-of-town centres.

The banking company sees its work space as divided between two key functions: the bank branches and the processing/call centre. There is a geographical separation here as well as a functional separation, with the branches in the high-street and the processing centre in an out-of-town industrial estate. Internally, space is organised in quite distinctive ways. Within the branch (in Manchester), the manager describes the bank as having three parts, 'the counter', 'out front' and 'out the back'. In the processing centre, there are large open rooms, with up to one hundred desks, separated into the different sectors of work tasks. In the branch the employees are expected to move between the three spaces, as dictated by the demands of the customers. In the processing centre the work-spaces are more tightly defined, separated according to job task. In terms of work practices, there is a clear division between the two work spaces. The staff in the branches are engaged in customer-related tasks and have 'sales' roles, while in the processing centre, they 'work round the clock on shifts . . . using high-tech equipment' (Bank branch manager, Manchester) to do routine clerical tasks. The branch staff are expected to be flexible and to self-manage their time, as the manager comments,

> The emphasis is on customer service . . . The important thing is that if we get a queue at the counter we must get more people on the counter or we must get someone to encourage people to use the self-service machines. Similarly if no-one's at the counter I would expect some of the cashiers to go round the back and ask for something to do. If lending had a lot on, I would expect someone to be flexible enough to go from the ground floor to the first floor to help out. The object is, that although someone maybe a 'cashier' or an 'adviser', if there is a bottleneck in the bank they've got to be flexible enough to now 'think customer, must do something'. That's what we're trying to get people to do.
>
> (Bank branch manager, Manchester)

So, job demarcations largely disappear in the branches of the bank, with all staff being expected to carry out most tasks. As the manager adds,

> In an ideal world everyone would be able to do everything. We're not close to that yet. We have some people who can do every job. What we need is a lot more of those sort of people.
>
> (ibid.)

There is a breaking down of job demarcation, combined with the expectation of multi-tasking and adaptability and also a shift in skills to information tech-nology. As McDowell argued above, although *all* work has *always* been embodied, many service occupations – including banking – are increasingly interactive and 'embodied' as staff are expected to be flexible, adaptable and to perform and 'become' their employment.

Jane was, until recently, an employee in the bank's Manchester high-street branch. She had been employed for several years working on the switchboard. She is severely visually impaired, but this caused her no problems in her job. In 1996 the company centralised all its telephone services to two regional sites, which were too far away for Jane to travel to. As a result she became a cashier in the branch, with special magnifiers to enlarge the computer screens. But, as the branch manager put it,

> To cut a long story short we persevered and she was very grateful for all we'd done, but it got to the point where she was stressed because she couldn't do what everybody else was having to do. It was agreed that she's leaving with a package. I feel disappointed because she's a lovely girl, [but] Jane was 50 per cent efficient, and on reception you need someone who can read the screens.
>
> (Bank branch manager, Manchester)

We can think about Jane's experience of employment through the two main explanations of disability and employment explained in the first section. The first explanation focused on the individual, placing the cause and solution with the disabled employee and the discriminatory employer. Jane's experience, in this interpretation, was due to her worsening impairment and her inability to adapt to the changing employment expectations of the bank, combined with the poor attitude of the employer (not necessarily the manager quoted) who failed to alter the workplace sufficiently for her to continue in employment. Thinking through the example in relation to the second explanation, we can see that the structures and processes of the capitalist work relation present in the bank's changing employment practice meant that Jane's abilities and acquired skills were devalued and her impairment became a disability.

However, neither of these understandings really gets to the heart of the matter. Jane's (embodied) experience of work is missing from the discussion. Jane had a job that she could do and was happy with, on the switchboard. The relocation of the company's telephone system to two call centres could be seen as indirect discrimination as the company did not make allowances for the travel that Jane would have to do. But the distances were such that most employees would not have wanted to travel. Many people with visual impairments *know* that they cannot travel long distances or *do not want to* because of the difficulties they face, however good the transport system. So, Jane took up her job in the branch as a cashier and the company provided access by enlarging the screens for her. But she began to find reading the screens difficult and this affected her work, something she was well aware of and unhappy with. The company was attempting to enable her to work and she was capable of her work. But her bodily involvement in the work – the use of her eyes, the interaction with the customers – was not how she wanted it to be. As the manager stated she was stressed because she could not do what other employees were having to do. So, Jane left the employment of the bank.

While the second explanation (and the social model that it echoes) provides a powerful political interpretation of the overall position of disabled people in employment, it is less useful in explaining the everyday employment experiences of someone like Jane. As the above sections have argued, the theory of the body is useful for rethinking the disability–employment relation. Jane's employment involved her whole body. There are three issues here. First, she was closely involved in her different employment tasks as they interacted with her body, which in turn reacted back into the employment. The necessarily embodied performance of her employment, within the defined codes of work, meant that she 'became' the job (Pinder 1995; McDowell 1994). The oppositional nature of both main explanations is less useful as it separates the disabled person and employment. Second, impairment *does* matter. Recognising that Jane's visual impairment does have an impact on her employment (not necessarily negatively), while still maintaining the underlying structure of the oppressive capitalist economy (and the social model) is a significant development in thinking about disabled people and employment (Crow 1996). Third, there is another important issue in Jane's story and her decision to leave. Both explanations of the disability–employment relation leave disabled people largely outside the debate. French argues that, 'is it not time for people with impairments to decide whether or not the problem they experience amount to disabilities?' (1993: 19). We could understand Jane's experience as a common sense response to employment and her impairment. Jane took the decision that she did not want to stay within the company for the sake of it and do an alternative job that did not use her skills. So she decided to leave, because she could not do the job she wanted to do, despite her own perseverance and the efforts of the bank branch manager. This is recognition of the experience of employment common to many disabled people.

It is worth stating that an embodied approach also allows us to think about the category of 'disability' in employment and as a consequence the category of 'ability'. Jane is not the only employee in Central Bank experiencing the changes in employment expectations and work and spatial restructuring. Blurring the edges of who is disabled and who is able-bodied gets us away from the oppositional thinking that dominates the disability–employment debate.

The spatial organisation of employment is central to this rethinking. The bank, as noted above, divided its work space between branches and telephone/processing centres. For Jane, the transfer of telephone services from the branch to a processing centre was a significant part of her changing her type of employment. But, importantly, it was not a simple case of her being spatially excluded from the job. Jane actually decided that she did not want to move to the processing centre because it would be too far to travel. The spatial reorganisation of the company did raise issues of accessibility and exclusion in relation to disability, but the experience of a disabled employee involves many other issues of impairment, lifestyle and personal choice. Her experience was not solely determined by the decision-making of the company.

Second, the expectation of the company that all its employees are flexible and adaptable in terms of job tasks and within the work space could be seen to have excluded Jane from her position as a cashier. However, as above, the situation is a little more complex than that. The connection between the employee's body and their work is much closer (as the body and work interact) than managers may believe and so movement between tasks and spaces is not as fluid and easy as is suggested (and indeed most managers actually realise this). Space is in fact quite rigidly structured in a company, as the split between the branches and processing centres showed. So, Jane's decision to leave her job at the bank was not a direct result of the flexible capitalist work relation forcing her out of job, but rather again a complex mix of personal experiences and decisions about work and 'ability'. What is important here is the way that employment roles and relations are *negotiated* in the work spaces, as disabled (and able-bodied) employees work out their position and identity through their bodies and experiences.

Refiguring disability and employment

It is now important that we pause and think about how we can perhaps refigure the relationship between disability and employment. The above discussion and case study have shown that there is a different way of looking at the relationship, one that attempts to capture the complexity of the positions and experiences of disabled people in employment while still recognising the oppression and discrimination that disabled people face. The discussion of the case study showed that there are three key issues: first, that employment has real effects on the employee's body and the body then has real effects on employment; second, that these interactions and experiences take place within a framework of rules, codes and performance about what bodies are acceptable and which are not; third, employment operates within certain work spaces and employees 'work out' their position and identity within these spaces. We can think of these three issues together in the sense that the processes of employment produce material effects on people in employment, and that the nature and extent of these processes and the way they are understood is governed by a framework of meanings and codes, operating within spatial structures.

French describes this complex situation, using her experiences of lecturing in a college. Her difficulty in reading non-verbal clues *is* an issue,

> It is true that to some extent non-verbal communication can be replaced
> by verbal communication, but in reality the subtleties of non-verbal com-

munication are difficult (perhaps impossible) to replace; a student may look bored, or interested, but is unlikely to verbalise such feelings. One social solution to these difficulties would be for me to give up teaching large groups of students altogether, or to have a sighted colleague with me all the time; all in all, however, the lectures were successful, the students were satisfied, I was sufficiently familiar with the situation to cope with it, and in many ways the problem, though far from trivial, was insufficiently serious to warrant any drastic action. The situation I have described is not concerned solely with visual impairment, for it involves social interaction, but neither is it born of social oppression.

(1993: 19)

French elsewhere describes this as a 'middle ground'. I would argue that her experience of a 'middle ground' of impairment and disability within a social and bodily world is the everyday life of most disabled people.

Concluding thoughts

The experience of employment for a person with an impairment can be summed up in the following way: 'External disabling barriers may create social and economic disadvantage but our subjective experience of our bodies is also an integral part of our everyday reality' (Crow 1996: 210). And she proposes that,

> We need to find a way to integrate impairment into our whole experience and sense of ourselves for the sake of our own physical and emotional well-being, and subsequently, for our individual and collective capacity to work against disability.
>
> (ibid.)

If we think of employment and disability as *embodied* experiences, then we can include the issue of impairment and so possibly provide a fuller understanding of disability.

The 'disability' label, as argued above, is a social construction, dependent on the social and cultural conditions of the time. So, what is meant and understood by 'disability' changes over time. But the disability identity is an embodied experience, something which, although changing with social conditions, also carries, or embodies, physical and mental experiences about what is possible and what is not possible. If 'disability' is an unstable category, then it can be argued that 'ability' is also unstable and socially constructed.

So what we have is a pair of shifting labels (disability and ability) that are understood and used by government, employers, and society, and at the same time related identities, which are themselves changing but in a much slower and more complex way. The government's 'Welfare to Work' proposals reinforce the notion of the clear divide between ability and disability, so missing the wide range of issues that face disabled people in work, as the chapter has outlined.

This chapter has attempted to disrupt the conventional understandings of the disability–employment relationship by looking at the changing patterns and spaces of work in one company. It is most definitely not saying that discrimination against disabled people, despite the Disability Discrimination Act (1995), does not exist – it most definitely does. But we have to be clear that the experience of employment, and the discrimination within that, is highly variable and is much more complex than commonly thought. Disabled people in employment are, like all employees, capable of and limited in, by their bodies, certain types of work. It is the relationship between the body and work in space which lies at the heart of the disability–employment relationship.

Acknowledgements

Thanks go to Hester and Ruth for setting up the session at the RGS–IBG conference in January 1997 which started this whole project, the excellent comments, and their determination and skill to get this book published. Their patience too has been phenomenal. The constructive criticism of two referees and Gordon Hughes helped enormously. Thanks also to the staff of the banking company, who necessarily must remain nameless. The research was funded by an ESRC Research Studentship Award, R00429434141.

Notes

1   The recent *Labour Force Survey* (1997) showed that 26 per cent of disabled employees were in semi-skilled or unskilled jobs, compared with 20 per cent of all employees. Four per cent of disabled employees are in professional jobs, compared with 6 per cent of all employees. On earnings, the SCPR (Prescott-Clarke 1990) survey showed that of disabled men in full-time employment, 13 per cent received less than £100 a week, while 2 per cent of all men received the same amount. At higher levels of pay, 5 per cent of disabled men earned weekly over £400 pounds, compared with 12 per cent of all men. The *Labour Force Survey* also revealed an unemployment rate amongst disabled people of 20 per cent, compared with 7.5 per cent for all people seeking work.
2   All names have been changed.

References

Barnes, C. (1991) *Disabled People and Discrimination in Britain: A Case for Anti-Discrimination Legislation*, London: Hurst in association with British Council of Organisations of Disabled People.

Bell, D. and Valentine, G. (1995) *Mapping Desire*, London: Routledge.

Berthoud, R., Lakey, J. and McKay, S. (1993) *The Economic Problems of Disabled People*, London: Policy Studies Institute.

Bury, M. (1996) 'Defining and researching disability: challenges and responses', in C. Barnes and G. Mercer (eds) *Exploring the Divide: Illness and Disability*, Leeds: The Disability Press.

Crang, P. (1994) 'It's showtime: on the workplace geographies of display in a restaurant in south-east England', *Environment and Planning D: Society and Space* 112: 675–704.

Crow, L. (1996) 'Including all our lives: renewing the social model of disability', in J. Morris (ed.) *Encounters with Strangers: Feminism and Disability*, London: The Women's Press.

Disability Discrimination Act (1995) London: HMSO.

Finkelstein, V. (1981) 'Disability and the helper/helped relationship: an historical view', in A. Brechin, P. Liddiard and J. Swain (eds) *Handicap in a Social World*, London: Hodder and Stoughton in association with The Open University Press.

Foucault, M. (1974) *The Archaeology of Knowledge*, London: Tavistock.

French, S. (1993) 'Disability, impairment or somewhere in-between?', in J. Swain, V. Finkelstein, S. French and M. Oliver (eds) *Disabling Barriers – Enabling Environments*, London: Sage.

Gleeson, B. (1996) 'A geography for disabled people?', *Transactions of the Institute of British Geographers* 21: 387–96.

Graham, P., Jordan, A. and Lamb, B. (1990) *An Equal Chance? Or No Chance?: A Study of Discrimination Against Disabled People in the Labour Market*, London: The Spastics Society.

Grosz, E. (1994) *Volatile Bodies*, Indiana: Indiana University Press.

Hochschild, A. (1983) *The Managed Heart: The Commercialisation of Feeling*, Los Angeles: University of California Press.

Imrie, R. (1996) *Disability and the City: International Perspectives*, London: Paul Chapman.

Jordan, W. (1982) 'First impressions: initial English confrontations with Africans', in C. Husband (ed.) *Race in Britain*, London: Hutchinson.

Kaplan, G. and Rogers, L. (1990) 'The definition of male and female: biological reductionism and the sanctions of normality', in S. Gunew (ed.) *Feminist Knowledge, Critique and Construct*, London: Routledge.

Labour Force Survey (1997) Winter 1995/6.

Labour Research (1997) 'Victory in first disability cases', October.

Lee, R. and Wills, J. (eds) (1997) *Geographies of Economies*, London: Arnold.

Longhurst, R. (1995) 'The body and geography', *Gender, Place and Culture* 2: 97–105.

McDowell, L. (1993) 'Power and masculinity in city work places', paper presented at the Institute of British Geographers Annual Conference, London.

—— (1994) 'Performing work: bodily representations in merchant banks', *Environment and Planning D: Society and Space* 12: 727–50.

Martin, J., White, A. and Meltzer, H. (1989) *Disabled Adults: Services, Transport and Employment*, London: HMSO.

Morrell, J. (1990) *The Employment of People with Disabilities: Research into the Policies and Practices of Employers*, London: HMSO.

Morris, J. (1991) *Pride Against Prejudice: Transforming Attitudes to Disability*, London: The Women's Press.

Oliver, M. (1990) *The Politics of Disablement*, London: Macmillan.

Pile, S. and Thrift, N. (eds) (1995) *Mapping the Subject*, London: Routledge.

Pinder, R. (1995) 'Bringing the body back without the blame?: the experience of old and disabled people at work', *Sociology of Health and Illness* 17: 605–31.

Prescott-Clarke, P. (1990) *Employment and Handicap*, London: Social and Community Planning Research.

Rose, G. (1993) *Feminism and Geography*, London: Polity.

Rose, S. (1997) *Lifelines: Biology, Freedom, Determinism*, London: Penguin.

Ryan, J. and Thomas, F. (1980) *The Politics of Mental Handicap*, Harmondsworth: Penguin.

Shilling, C. (1993) *The Body and Social Theory*, London: Sage.

Stone, D. (1985) *The Disabled State*, London: Macmillan.

The Guardian (1996) 'Banks "planning new wave of job cuts"', 13 September.

—— (1997) 'Welfare reform: how the Labour team fell out of step', 21 October.

Turner, B. (1992) *The Body and Society*, London: Sage.

8

AUTOBIOGRAPHICAL NOTES ON CHRONIC ILLNESS

Pamela Moss

Writing autobiographically

Sometime around March 1995 I fell ill. I developed a pattern of wicked headaches, terrifying nightmares, and crushing fatigue, the kind where your chest hurts and you are so tired that you do not have the energy to breathe. In my workplace, I had endured a horrific year, filled with harassment from colleagues and a series of professional betrayals. I was finding it difficult to concentrate on reading, to remember friends' phone numbers, and to keep my balance while walking. I thought that if I had some rest over the summer, I would improve. For years, I had been struggling with pain, and I thought I was just going through a 'down' period.

When teaching started again in September, I was overwhelmed with a debilitating fatigue. I needed to lie down for an hour after every little exertion: showering, eating breakfast, taking the bus to university. For an 11:30 lecture, I had to get up at 7:30, after which I spent the rest of the afternoon, plus the entire next day, in bed. Biomedical and naturopathic physicians immediately diagnosed *myalgic encephalomyelitis* (ME). I took the next six months off, holding on to only some academic duties, supervising graduate students and organising an international feminist conference of praxis and theory.

Chronic illness had been hovering at the edge of my life since I was in high school. But, the symptoms I suffered never interfered with any of the activities I was involved in. From that September until now, my life has been, and continues to be, organised around my illness, from what I eat to when I sleep, from where I go to whom I interact with, from what I think about to how I expend my energy. The ME permeates who I am and everything I do. Given my intellectual interest in chronic illness, for I had secured funding for a project researching the daily lives of women with arthritis or ME,[1] I wanted to use my experiences to assist me in building a radical body politics, to interpret how women restructure their environments in order to accommodate a disabling chronic illness.

But I found that writing about one's self is both daunting and invigorating: daunting in the sense that in weaving my *self* through the threads of other people's lives, it is difficult to sort through what is important, what is muck, and what is fluff; yet invigorating in the sense that in placing my *self* more widely, new connections emerge through which I can forge links with other people, other places. I already knew that this type of writing takes several forms in geography, termed in palatable phrases such as reflections, recollections, positionings, and personal writing (e.g., Billinge *et al.* 1984; Nast 1994; McGee 1995; Moss 1995; Tuan 1998; Valentine 1998), but rarely as autobiographical writing (for an exception see Moss forthcoming). I found in my readings that geographers persistently use various types of autobiographical writing to draw attention to lacunae in the literature, justify claims to authority, and develop theoretical tensions in the discipline (e.g., Chouinard and Grant 1995; Soja 1996; Monk 1997). In addition, an author commonly uses autobiographical anecdotes as a way to exude warmth and personability in an attempt to make the reader more receptive to what follows, after which any personal account vanishes (Miller 1991).[2]

For my journey through the autobiographical in this chapter, I chose to use autobiographical writing in two ways. First, I use autobiographical writing to recount in empirical terms the politicisation of my own experience of ME, of chronic illness. Second, I use autobiographical writing to highlight the ways in which an analysis of these same experiences contributes to a general framework for theory formation on and of the body. My journey, at times, is ambiguously unfamiliar, and at others, lucidly intimate. This emergent discrepancy in presentation has more to do with my daily life with chronic illness, and less to do (at least, I think) with writing style, analytical insight or conceptual clarity. I first place my casting of autobiographical writing in the feminist literature. I then develop four empirical themes giving rise to discussion about theory. I conclude with comments the usefulness of engaging in autobiographical writing.

Placing autobiographical writing in feminist geography

Interactive processes between the researcher and the 'researched' have been discussed at length by feminist geographers (*The Canadian Geographer* 1993; *The Professional Geographer* 1994, 1995; *Antipode* 1995). Discussion centres on the two themes of reflexivity and positionality. *Reflexivity* in the research process involves an active, conscientious contemplation of how the researcher fits into the predominant power structures in society, polity, and economy, particularly in comparison to the women who are participating in the research project. Once positioned, the researcher acts as mediator, translator, or supplicant of the women's stories, which then represent the women's lives in text. Through this *positionality*, the researcher situates her interpretation of other women's lives *vis-à-vis* the reader, who can conceive the women's daily lives more readily. It is through these reflections that the researcher can come to a fuller understanding and perhaps can construct better explanations of the phenomenon under study.

Through an autobiographical writing that is rooted in the task of reflexivity, I can make explicit the process of positioning one's *self*, through exposing some of the rich detail needed to make sense of both our own and others' daily lives. In this way, autobiographical writing is different from using one's autobiography merely as an additional source of information, as for example in adding my story of chronic illness to the stories of the women Kathleen,[3] Isabel, and I talked with in the chronic illness study. It also differs from seeking out experiential knowledge through complete participant observation, as I did in a previous study with franchise housekeepers in a city in Southern Ontario, Canada (see Moss 1995), because there is more reflection on and scrutiny of my own experiences. Such autobiographical writing is also more than a confessional or self-analysis (through many of the forms of psychoanalysis) because I *reconnect* my insights from my own reflections to the wider context within which I exist for purposes other than self-understanding. Similarly, through autobiographical writing, I can move beyond reciting my self history or self story, towards an informed reading of politicised experience. Although hunches can provide direction in a research project, autobiographical writing is different from exalting intuition in that claims about women's lives or about a research topic can be demonstrated, not just sensed. In addition, in contrast to Denzin (1989), who focuses on key events, or epiphanies, I suggest that it is just as beneficial to emphasise routines and mundane acts of daily living, the stuff comprising our everyday life that makes us think as we think, believe what we believe, and act how we act. It is in all these senses that the autobiographical writing I am venturing into involves more than just reflexively positioning *self* in the research process for it can bring to light how *self* is positioned in multiple milieus within a framework for interpretation and explanation. Through a journey into the autobiographical, *self* can become integrated into the interpretation and the researcher's process of interpretation and explanation can, in addition to the phenomenon under study, come under scrutiny.[4]

Like most other interpretative types of analysis, autobiographical writing is not an end in itself, nor does the process ever cease. It is both a data collection and an analytical method, sensitive to the reflexive nature of positionality in theorising processes of oppression, marginalisation, appropriation, and colonisation. We all must begin someplace, and that place for me, right now, is here. So, in what follows, I use the discussion thus far to frame my journey through the autobiographical, through my experience of ME, a chronic illness.

Some autobiographical notes on chronic illness

Below are four themes which demonstrate how my illness in its early stages can be used to articulate my thoughts on research and theory building. The first two themes deal primarily with using autobiographical writing as a way to accent the process of awareness, and the subsequent politicisation, of my own experiences of ME. The latter two themes demonstrate how autobiographical writing can contribute to framing the construction of theory.

Conception and the 'gap'

During the design of the project, I drew on my experiences with a multi-diagnosed illness. Because I had gone through years of pain, restlessness, and insomnia, I thought that I would be able to empathise with the women I would talk with in the study, both in terms of symptomatology and what it is like to live with a chronic illness at home and in the workplace (e.g., Pinder 1988, 1996; Charmaz 1991). Although I had not been diagnosed with ME,[5] I had been diagnosed with several other things including rheumatoid arthritis (RA), schleraderma, relationship stress, and fibromyalgia (FM). FM was the diagnosis that I latched onto, and for the most part still do consider it part of my illness, because it explained my medical history and the wide range of health problems I was experiencing.

I initially sought common ground upon which to build links within the research process and participants – the symptoms, the diagnostic process, and the bio-medical label. As part of contacting women to be part of the project, I planned to be sensitive to the way women are multiply positioning within various constellations of power. In the process of my own positioning, I found myself listing pages of adjectives, as for example, high-income,[6] white, female, ill, vegetarian-who-wants-to-be-a-vegan, and a 'wannabe' blonde. But these made no sense without the substance of everyday living that would pull these positions together into a life, my life. I cannot tell you everything about me, or even if I could I do not think that this would make sense given that I cannot lay bare my positioning with regard to the women in the study. Isabel, too, has her own positionings to work through, not the least of which is her relationship to the women as someone without chronic illness.

Empathy and theory

There is no question that having ME and FM helps me understand the women's experiences of chronic illness. When a woman talks about the difficulty she has in getting her friends to realise that she simply cannot *plan* to go out, I know what she means. Like other women, I have lost friends because I could not commit to a pre-arranged time, or if I did, in breaking engagements too often. When a woman talks about people not being able to 'see' her pain unless there

158

are splints, canes, or wheelchairs, I know what she means. No one 'sees' my disease either, so the dread over having to stand up on the bus for twenty minutes, the stopping, the going, the swaying, overwhelms me so much that I break out in a cold sweat at the thought of it. When a woman says she has a good day and ends up cleaning out a closet, but 'pays' for it the next three days from being so stiff she can hardly get out of bed, I know what she means. On a good day I have stayed up too late, walked too far, or socialised too much only to be reminded over the next few days when I am lying on the couch nodding off and wishing that someone would bring me a cup of tea.

Even though my empathy does not in itself contribute to theory, it does guide me in a promising direction to find information that will assist in building theory critically. It is in the write-up of research results where the analysis exposes and discloses the material and discursive links of autobiographical writing in its context. But it is more than just empathy that I should bring to the research process, to these women's lives, if I am going to use my own experiences analytically. I should bring with me an entangled mass of experiences and ways of knowing and being that have been heightened sensually because of my illness.

Inscription and disease

By inscription I mean a set of concrete processes, including both discursive and material practices, that etch onto the body a particular rendering of a disease category. The material, as both concrete and economic, is not always separable from the discursive, as ideological and linguistic, nor vice versa. They intersect and operate through a set of assumptions upon which individuals, existing within the structural ensembles of, for example, biomedical, insurance, legal, university, and governmental communities, base their interaction with an individual body inscribed with a particular disease category. Inscriptions of disease categories work in a variety of ways – with a specified disease category, there is a set of assumptions working to define who you are; without the category, the process is the same, the set of assumptions is different.

My inscribed body is both a site of oppression and a site of resistance. I have been experiencing some ME symptoms since 1987[7] and FM symptoms since I was in high school. Over the years I have consulted numerous physicians, general practitioners and specialists, within the biomedical community and received, in addition to the variety of diagnoses, myriad pharmaceutical-based treatment regimes, including non-steroid anti-inflammatory drugs, anti-depressants, birth control pills, tranquillisers, amphetamines, and anti-malarial drugs. None were useful and I am happy to say that I was able to reject them all, but not without some struggle. Other, less destructive treatments, I found more useful, as for example, physio-, hydro-, occupational, and massage therapies, but they did not make the pain go away or stave off fatigue. What is interesting about this inscription is that I was able to resist some parts but not all: I came to think of my physical symptoms as 'unreal', not that they were not there, but they were not real *enough* to be taken seriously by someone who supposedly knew about pain and fatigue.

The history of my bodily inscriptions influences my capacity and the parameters within which I resist. Two different physicians diagnosed RA, once when I was 15, and again at 29. My relative youth in the face of a degenerative autoimmune disease had family members and friends dwelling on my condition, discussing imminent limits on my mobility, and planning joint replacements. I earnestly resisted this inscription by dealing with the pain so that I could play basketball at 15 and work on a Ph.D. at 29. When I was younger, I more willingly accepted an expert's knowledge with parental support about illness and disease. Later, when I was able to access biomedical knowledge on my own, I was more reluctant to accept at face value what any physician maintained. This approach led me to prompt the rheumatologist to consider FM as a possible diagnosis.

Once FM became the official diagnostic label, my bodily inscription notably shifted.[8] Socially, I increasingly isolated myself because I could not commit to any event in advance. In my workplace, more people assumed I was stressed from my job and that my physiological symptoms were psychosomatic. This inscription had contradictory implications. On the one hand, the administration took the disease category as authentic and provided equipment to assist me in creating an environment where I could work without exacerbating my symptoms. On the other hand, some members[9] of the same administration used my diagnosis to dismiss my claims of personal harassment by a colleague, by stating that I was unable to handle stress which resulted in my overreacting to mild taunts by and personality traits inherent in the colleague. In the first instance, the administration identified me as a researcher and teacher who needed some assistance in undertaking the tasks associated with my job. In the second instance, when the inscription threatened the advantageous positioning of masculine dominance in academia generally, the inscription was used negatively against me and I was identified as a typical female who is *too* emotional.[10]

With my diagnosis of ME in September 1995, I have gone on medical leave for six months and have returned to university with a reduced workload.[11] As an inscription, ME was far more complicated than FM. An ME diagnosis legitimated my health problems for my friends because it gave them an experienced-based explanation of my symptoms: most had heard of someone who was ill with ME. In contrast, in the workplace, the inscription fragmented and followed several different paths. Many students were sensitive and offered help, ranging from grocery shopping to suggesting names of sympathetic physicians. For some colleagues, ME was a manifestation of stress which fully demonstrated to them that I could not handle my job. For other colleagues, my inscription signalled the end of *my* career and the beginning of *their* more sensible approach to academic life. For the most part, I resist the negative connotations of this complex inscription by being open about ME and the status of my recovery to students, colleagues, and administration members. But entering into dialogue with these competing readings only exacerbates my symptoms and places me in a position where any decision I make about my life either at home or in the workplace challenges my experiences of being ill.

My struggle with the administration in my university over my claim for partial long-term disability insurance demonstrates the complexity of inscriptions vying for dominance at any one conjunction. Conversations with my claims analyst with the insurance company led me to believe that, circumstances permitting, partial long-term disability insurance benefits could be paid out for people like me – one who can work full time with a restructured workload. When I broached this possibility with administration, I was rebuked and told that *all* such arrangements have been made *only* when the person was declared *fully* unfit to work at all, a category with which I was not designated in the insurance assessment. Whether or not this is the case is not the issue; what is important here is the reliance on the specification of parameters that makes one 'disabled', parameters that leave persons with debilitating chronic illness at a material loss because of their fluctuating symptoms. In the case of ME, particularly, the marginalisation of the disease category in some biomedical communities has repercussions on my bodily inscription. For example, when the claims analyst submitted her report, she noted that my illness was probably related to my workplace environment. Because the insurance company turned down the claim, however, administration 'read' the report as simply saying that I was not ill. When I objected, the argument shifted to one that challenged my commitment to 'getting better'; citing that I refused to take pharmaceuticals.[12] Again, once I objected on the grounds that I had every right to pursue alternative medicine, the argument again shifted, drawing on my choice to quit work completely to recover. In cultivating such a complex response to the insurance company's denial of benefits, the administration was using the diagnostic label of chronic illness in multiple ways: first to define me as not ill, then to cast doubts on my commitment to strive for health. Both deny my experience of illness, and inscribe my body discursively and materially as *deviant*, one that is neither good at being ill nor good at being healthy. My insight into this complex appropriation of the authority gained from insurance companies in the doling out of disability benefits sets me in good stead to understand the material struggles women with chronic illness go through, those who are neither fully abled nor fully disabled.

My interest in ME and RA is a direct result of biomedical discourse inscribing these disease categories onto my body as well as the appropriation of the diagnosis and legitimacy by other discourses in order to explain why I feel so shitty all the time, both physically and in my workplace environment. My experiences of resisting hegemonic inscriptions of what it is like to have a chronic illness socially and in the workplace has informed my thinking about the body as a site of both oppression and resistance in mine and Isabel's contribution to the construction of a radical body politics (see Moss and Dyck 1996, 1999).

Resistance and body

My resistance is not only to biomedical inscriptions, but also to certain explanations of my experiences of my body. My experiences do not fit theories of the body as they have been predominantly portrayed, especially in feminist psychoanalysis. And, in contrast to those approaches to the body that Pile and Thrift (1995) identify, that is, logical, prediscursive, psychological, cultural, and social, my body is more like Grosz's (1994) conception of the corporeal body where the material, the discursive, and the biological are simultaneously encapsulated. Because of the unpredictable course of most chronic illness, flexibility and adaptability are necessary on the part of the ill individual (body), of individuals associated with the ill (bodies), and of the spaces through which individuals (ill bodies) are constituted as ill, healthy, recovered, or deviant. Because the boundaries of corporeal bodily experiences are fluid and constantly in flux, it is here, in the context of the biological as physiological, that the simultaneity and separability of discursive and material practices are most obvious. Corporeal experiences shape the spaces that our bodies traverse and impact the way other spaces, such as biomedical spaces, define our experiences hegemonically.

Yet I do not think, nor would I suggest, that this is the only way to experience the body. Resisting theoretical orthodoxy is imperative if I take seriously the claims of partiality in knowing and knowledge. Wolff's (1995) work, although not directly about the body, is useful in identifying a path that theorises both general culture (body) and specific field studies (bodies with chronic illness). She maintains that eclectic theory necessarily emerges when linking together disparate cultural moments (an ill and health body), if based on personal encounters with particular cultural phenomena (disease). She suggests that multiple experiences of the same cultural phenomenon might also produce eclectic theory. As a result, '[t]he exploration of a cultural moment through the experiences of particular women's lives will not produce a single coherent narrative, but rather a series of distinct but overlapping perceptions' (Wolff 1995: 35).

That cultural moment here, in this chapter, in my life, and in my research, comprises the multiple discursive and material inscriptions of ME and RA on

162

women's bodies as sites of oppression and of resistance. Parallel to Wolff's argument, a juxtapositioning of inscriptions of ME and FM will not produce a coherent narrative; however, there will be overlapping experiences, experiences that need not be theorised into a coherent series of abstract postulates, but rather more manageably into a (partial) explanation, as in, for example, the specificity of particular processes involved in the constitution of an ill identity by inscribing a disease category onto a body through material and discursive practices. Autobiographical writing about body and bodily inscriptions can and should lend itself to destabilising the construction of feminist theory of the body; not as an new orthodoxy, but rather as a series of analytical narratives relating concrete specificities to collective principles.

Concluding remarks

Casting autobiographical writing in the terms I have suggested – detailed accounts of the politicisation of experience in association with framing theoretical contributions – reorients the use of autobiographical writing away from merely being a source of information, a self story, an exercise in self awareness, and even a political catalyst towards a basis for analysis, into a framework for scrutinising, examining, and exploring social phenomena. This type of autobiographical writing entails thinking through how one can ally experience and critical thought in such a way as to produce a well-informed piece of research that is not taken out of context, does not create a false distinction between structure and process, and does not break the living connections in our own lives. This casting also continues the work of feminist geographers in their interest in reflexivity and positionality. Using autobiographical writing is more than positioning oneself in the research process. Self-reflection coupled with engaging experience leads to an entangled relationship with my *self* and to other women involved in my research with an expectation of a fuller understanding about a specific phenomenon, such as chronic illness.

This intersection between cultural and medical studies needs to be developed further in the feminist geographic literature, and not to be isolated and categorised only in medical or disability geographies. Specific issues arising around inscriptions, and resistance to them, further the discussion of feminist theories of the body, especially with respect to how processes of embodiment are imbued in particular places, as for example, the workplace. But it should be clear that my advocacy of autobiographical writing is not for everyone or every analysis. However, if one uses chooses to write autobiographically, then there has to be a recognition that experiences are embodied in context. With this recognition also comes the chance to spatialise theories of the body and bring them closer to our own experiences.

Postscript, from autumn 1998

I drafted this chapter nearly two and a half years ago. So much has changed since I first put my fingers on the keyboard to record and use for analysis some of my experiences of being chronically ill. My day-to-day life is still circumscribed by my illness and the demands people make on me proceed unabated. And I am continually struck by the dichotomous reactions of colleagues and students, either compassion and understanding or dismissal and coercion. I sometimes think about giving in to ME and FM, and permit them to take over my life, particularly on the days where untenable, non-discriminating remarks are made about me preferring a life of illness over one of health, being a disabled (read unable) academic, or having affected (read skewed) scholarship. But I do not. I persevere, but not in any kind of martyred way. I persevere for two reasons. First, more personally, I cannot stand the thought, let alone the bodily sensation of yet another headache, yet another nightmare, yet another day of complete, utter, debilitating fatigue. Second, it is vital for me to assist in politicising chronic illness as a disabling condition so that there is no dismissal based on otherness and no coercion to conform.

Acknowledgements

I thank Hester Parr and Ruth Butler for their critical comments which helped me focus my discussion and set my arguments into a more readable format. I also thank Michael Dorn for his helpful suggestions. Karl bore the brunt of my anxiety over finalising the details in this paper – I have to thank him, yet again, for his patience and support.

Notes

1 The project is funded by the Social Science and Humanities Research Council of Canada (SSHRC), #410 95 0267. Isabel Dyck is co-investigator. We focus on the everyday activities of forty-nine women diagnosed with either ME or rheumatoid arthritis (RA) and how they structure and restructure their physical and social environments. The project is not yet complete.

2 There are types of autobiography which are part of academic writing, particularly in feminist scholarship. For example, as a literary genre, autobiography has undergone considerable redefinition in the past two decades, especially with the development of feminist literary criticism (e.g., Jelenik 1980; Stanton 1984; Greene and Kahn 1993). Further, delineation of what comprises feminist autobiography is at issue as the boundaries between biography and autobiography as well as those between truth-correspondence and fiction are being re-worked (e.g., Stanley 1992; Brossard 1995). In this chapter, I do not address any of these issues.

3 Kathleen Gabelmann, a graduate student and research advisor, undertook about half of the interviews in Victoria and most of the ones in Vancouver.

4 I think that this type of autobiography is most clearly articulated in lesbian theory (see, for example, articles in Hart and Phelan (1993) and Domenici and Lesser (1996)).

5 Only after the project was funded was I diagnosed with ME.

6 I am high-income in relation to women and middle-income in relation to men.

7 Biomedically, this statement is untrue. The characteristic debilitating fatigue of ME came over me at the end of March 1995.

8 The diagnosis of schleraderma lasted only four days, also when I was 29. I told only two people, so I did not experience the material practices that tend to reinforce hegemonic discursive inscriptions. It is difficult to remember what exactly I was thinking when I was tagged with a terminal illness with an expected life span of about five years, as it was at that time. Though short, I remember its power to force me to think about what I really wanted out of life. Then, after the doctor told me it was the wrong diagnosis, its incription quickly dissipated with only a remnant remaining: I needed a physiological reason for the pain.

9 There were some members of the administration who took my claims seriously and acted accordingly. A professionalism I still appreciate.

10 Here I would argue that 'too' is extremely important because men in the 1990s are permitted and even encouraged to be emotional, but not *too* emotional like women.

11 As I revise this chapter, I am considered 'full-time' on sabbatical, but still suffering from ME.

12 There is not one course of action to treat ME. Although pharmaceuticals are prescribed to some ME patients, physicians are specific in saying that such drugs do not treat ME; they treat only some of the symptoms.

References

Antipode (1995) 'Discussion and debate: symposium on feminist participatory research', *Antipode* 27, 1: 71–101.

Billinge, M., Gregory, D. and Martin, R. L. (eds) (1984) *Recollections of a Revolution*, London: Macmillan.

Brossard, N. (1995) 'Green light of Labyrinth Park, La Nuit Verte du Parque Labyrinthe', in E. Grosz and E. Probyn (eds) *Sexy Bodies: The Strange Carnalities of Feminism*, New York: Routledge.

Charmaz, K. (1991) *Good Days, Bad Days: The Self in Chronic Illness and Time*, Berkeley: University of California Press.

Chouinard, V. and Grant, A. (1995) 'On not being anywhere near the "project": revolutionary ways of putting ourselves in the picture', *Antipode* 27, 2: 137–66.

Denzin, N. (1989) *Interpretive Biography*, London: Sage.

Domenici, T. and Lesser, R. C. (eds) (1996) *Disorienting Sexuality: Psychoanalytic Reappraisals of Sexual Identities,* London: Routledge.

Greene, G. and Kahn, C. (eds) (1993) *Changing Subjects: The Making of Feminist Literary Criticism*, London: Routledge.

Grosz, E. (1994) *Volatile Bodies: Toward a Corporeal Feminism*, Bloomington: Indiana University Press.

Hart, L. and Phelan, P. (1993) *Acting Out: Feminist Performances*, Ann Arbor, MI: University of Michigan Press.

Jelenik, E. C. (ed.) (1980) *Women's Autobiographies: Essays in Criticism*, Bloomington: Indiana University Press.

McGee, T. G. (1995) 'Eurocentrism and geography: reflections on Asian urbanization', in J. Crush (ed.) *Power of Development*, New York: Routledge.

Miller, N. K. (1991) *Getting Personal: Feminist Occasions and Other Autobiographical Acts*, London: Routledge.

Monk, J. (1997) 'Marginal notes on representation', in J. P. Jones, III, H. J. Nast and S. M. Roberts (eds) *Thresholds in Feminist Geography: Difference, Methodology, Representation*, Boulder, CO: Rowman and Littlefield.

Moss, P. (1995) 'Reflections on the 'gap' as part of the politics of research design', *Antipode* 27, 1: 82–90.

—— (ed.) (forthcoming 2000) *Autobiography Engaging: Geographers Writing Lives*, Syracuse: Syracuse University Press.

Moss, P. and Dyck, I. (1996) 'Inquiry into body and environment: women work and chronic illness', *Environment and Planning D: Society and Space* 14, 6: 737–53.

—— (1999) 'Journeying through ME: identity, body and women with chronic illness', in E. Teather (ed.) *Embodied Geographies*, London: Routledge.

Nast, H. J. (1994) 'Opening remarks on "Women in the Field"', *The Professional Geographer* 46, 1: 54–66.

Pile, S. and Thrift, N. (eds) (1995) *Mapping the Subject: Geographies of Cultural Transformation*, London: Routledge.

Pinder, R. (1988) 'Striking balances': living with Parkinson's Disease', in R. Anderson and M. Bury (eds) *Living with Chronic Illness: The Experience of Patients and their Families*, London: Hyman Unwin.

—— (1996) 'Sick-but-fit or fit-but-sick? Ambiguity and identity at the workplace', in C. Barnes and G. Mercer (eds) *Exploring the Divide: Illness and Disability*, Leeds: The Disability Press.

Soja, E. W. (1996) *Thirdspace: Journeys to Los Angeles and Other Real-and-Imagined Places*, Cambridge, MA: Blackwell.

Stanley, L. (1992) *The Auto/Biographical I/Eye*, Manchester: University of Manchester Press.

Stanton, D. (1984) 'Autogynography: is the subject different?', in D. Stanton (ed.) *The Female Autograph: Theory and Practice of Autobiography from the 10th to the 20th Century*, Chicago: University of Chicago Press.

The Canadian Geographer (1993) 'Feminism as method', 37, 1: 48–61.

The Professional Geographer (1994) 'Women in the field', 46, 1: 54–102.

—— (1995) 'Should women count?', 47, 4: 426–66.

Tuan, Y.-F. (1998) 'A life of learning (1998 Charles Homer Haskins Lecture)', occasional paper no. 42, New York: American Council of Learned Societies.

Valentine, G. (1998) '"Sticks and stones may break my bones": a personal geography of harassment', *Antipode* 30, 4: 305–32.

Wolff, J. (1995) *Resident Alien: Feminist Cultural Criticism*, London: Polity Press.

9

WHAT IT MEANS TO BE A MAN: THE BODY, MASCULINITIES, DISABILITY

Gill Valentine

Introduction

Within the social sciences the 1990s have been marked by an increased recognition of the significance of using case study research, autobiography and other forms of personal narrative in enabling us to understand and shed light on the processes which operate in our everyday lives. There is a long history of case study research within the social sciences (see, for example, Platt 1992), although this methodology has been understood to mean different things at different times. In a contemporary context, case studies are understood to be selected on theoretical grounds and to be designed to answer particular theoretical or policy questions (Yin 1984). '[T]he object of the analysis [in case study research] is not in fact "culture" or "society" of which the events might be considered samples, but rather the social processes which may be abstracted from the course of events analysed' (van Velsen 1967). Further the intention in adopting a case study approach is not to invoke statistical inference from the findings. Rather as Mitchell (1983: 207) argues 'the validity of the extrapolation [from the case study to a parent universe] depends not on the typicality or representativeness of the case but upon the cogency of the theoretical reasoning'.

By adopting a case study approach to examining individual's, or household's, living arrangements and decision-making processes, researchers are able to unpack the *complexity* and sometimes *contradictory* nature of everyday experiences and to set these experiences in *context*. The *detail* produced by such an approach also facilitates the *depth* of the research by making visible the negotiated and contested nature of everyday life. Finally, the repetitive and multi-method approach adopted in case study research can capture the *dynamics* of individual or household behaviour by revealing how participants' accounts of their lives may change or have changed over time, and by exposing differences between potential, usual and actual behaviour (Milburn 1995). As a result Mitchell argues that:

the rich detail which emerges from the intimate knowledge the analyst must acquire in a case study, if it is well conducted, provides the optimum conditions for the acquisition of those illuminating insights which make formerly opaque connections suddenly pellucid.

(1983: 207)

This chapter takes just such a case study approach to explore Paul's, a 31-year-old ex-miner, changing experience of embodiment following an accident in which he sustained spinal injuries, and the processes through which his individual narrative of the self has been negotiated and (re)produced as he has been located in a shared narrative of identity not of his own making – that of a 'disabled' person. The findings are based on multi-method case study work including: five stages of interviewing, a food consumption diary, a shared meal, and a photographic project as part of a larger Leverhulme Trust funded study on food, place and identity (see Bell and Valentine 1997; Valentine and Longstaff 1998; Valentine 1998, 1999).

Despite the recent sociological interest in the body, Oliver (1990, 1996) argues that social science research has tended to look at disability through a medical lens, while disability studies have not engaged with sociological work. In the light of such critiques, medical geography has recently begun to incorporate social theory into the sub-discipline (e.g., Dyck 1992; Kearns 1993; Dorn and Laws 1994) and a new body of theoretically informed work on geographies of disability (e.g., Butler and Bowlby 1997) has begun to emerge, leading to new attempts to explore the way identities, physical impairment and space intersect. As Dyck argues:

Close attention to the body in material context provides the potential for exploring the involvement of dominant discourses and power relations in the social construction of ideas about the body and identities, including that of the 'disabled body' and the implications for the experience of place.

(1995: 308)

This chapter aims to looks at the complex relationships which exist between hegemonic masculinities and disabled masculinities. It begins by exploring the concept of hegemonic working-class masculinity and considers the extent to which Paul's bodily performances prior to his accident located him within this shared narrative of identity. It then goes on to consider how Paul's accident located him within a new shared narrative of identity not of his own making – that of a disabled person and how this affected his ability to continue to reproduce his previous highly embodied working-class masculine identity; finally it considers how Paul has gradually reconstituted these twin shared narratives of identity, by producing his own narrative of the self, through renegotiating his gendered and (dis)abled bodily identity in complex ways.

Doing gender: hegemonic masculinities

Judith Butler (1990) has very famously argued that gender is not something one has, or is, but rather something that is done or performed. She writes that 'gender is the repeated stylisation of the body, a set of repeated acts within a highly rigid regulatory framework that congeal over time to produce the appearance of substance, of a natural sort of being' (Butler 1990: 33). While gender identity then is a complex, fluid and precarious business in which multiple positions are available, the promulgation of particular gender discourses by the State, the media, popular culture and so on have 'amounted to the conflation of a spectrum of possible gender practices into a constricted and constricting realm of acceptable male attitudes and behaviours; a complex and often de-bilitating self-regulation by males of their gender and, most especially of male difference' (O'Neill and Hird 1998: 6). These hegemonic, or dominant forms of masculinity are predicated on particular sets of bodily performances. As Morris (1991: 93) argues '[hegemonic masculinity] is inextricably bound with a celebration of strength, of perfect bodies. At the same time, to be masculine is not to be vulnerable. It is also linked to a celebration of youth and of taking bodily functions for granted'.

Prior to his accident Paul's gender and class identity were produced through his ability to repeatedly realise particular performative acts. Acts which were dependent upon the physical capacities of his material body. Connell (1995) has argued that certain kinds of manual work labouring, lifting, digging, carrying and so on are closely linked to some sense of bodily force in masculinity. While economic restructuring means that the emphasis on pure labouring has declined, the social meanings and relations of physical labour and bodily capacity to masculinity have not (Connell 1995). As a miner, Paul's bodily capacity and his ability to endure physical hardship were his economic assets and were crucial to his identity as a 'working-class man'. In the following quote he describes the harsh conditions underground where he worked for long hours in extreme temperatures, sometimes on his hands and knees, with few chances to take a break for food and drink and nowhere to rest or relax.

> You used to go down underground about six o'clock in morning, you didn't come up till two clock and er – it's not just down, I don't know how deep it was, say five, I don't know how deep it were really, on, shaft sort of thing, we used to go down and then you might have to go, just trying to explain it from here er – you might be working five, ten mile away from where you go down . . . so – so you've got to

take everything with you like . . . you'd be carrying everything 'cause you, you got sort of thing, you got to have it in a bag over your shoulder 'cause you've got your lamp on, you've got your lamp which is a big battery what you have on your belt, you got a, a – can't remember what they called it, it were a silver thing like this, it's like a mask if there were a gas leak, then all your tools. In some places – say it were like these doors, big fire doors they're like bigger than that [pointing to a door] and say I went through there, you could have to be working in shorts, some blokes were that hot, then you'd come back here [back through the doors] and you'd need a donkey jacket on, it were that cold . . . soon as you go down your hands are black anyway, you're climbing, you know lifting things and that, there's nowhere to wash your hands so when you're eating your sandwiches . . . you have to eat round your fingers sort of thing. Well if not like you're eating dirt and stuff. And er – there were never a certain time [to stop for breaks], they used – 'cause end of day, it were all about getting coal off face . . . you weren't having an hour's dinner break and you weren't bothered about that because you were getting paid for what you were producing. Down pit, you get a bonus on what you produce as well as your wage, so more you produce – more money you fetched home . . . it's awkward to explain like . . . on face, they had to work on their hands and knees, they were that low, same as size underneath this table – so they used to come out of there to eat . . . I mean you might be sat there having yours and they'd come and sit, and you just sat, yeah.

Connell (1983) argues that cults of physicality and sport (taught and informal) give clear ideal definitions of not only how a male body should look but also how it should operate. While women are valued because of their bodily appearance – men, he argues are expected to be able to exhibit bodily skill in terms of their competence to operate on space, or the objects in it, and to be a bodily force, in terms of their ability to occupy space. He writes:

What it means to be masculine is, quite literally, to embody force, to embody competence . . . Force and competence are, obviously enough, translations into the language of the body of the social relations which define men as holders of power, women as subordinate . . . And it is especially important in allowing this belief and the attendant practices to be sustained by men who in other social relations (notably class relations in employment) are personally powerless.

(1983: 27–8)

Paul's ability to embody force and competence through sport prior to his spinal injury was particularly important to his definition of himself as 'masculine' and to his social relations (with both other men and women) although as he describes below, he did not have the self-discipline to work on developing his strength and sporting prowess – rather he took his bodily force and competence for granted as he describes in this quote:

> I used to play a lot of sport when I were at school. Well I 'ad me accident, before that I used to play a lot of football. Football more than anything really. An' I never thought about it really. I remember going on a – joining one of these clubs when I were about 17 – you know, these fitness clubs sort of thing. And they put me on this – and the instructor put me on this a course – well it's a course – you know diet about what you could an' that and I found it 'ard to keep then, because when you're like 17–18 you like to go out for a drink wi' lads and it's a bit 'ard to do. I used to think – used to go training like an' then come out wi' lads who'd been there then go and have like a couple of pints and go to chip shop so that were all wrong? But I never used to think much about it then, at that age . . . like any other person of that age I didn't really think.

Of course, what it means to be masculine is not only to embody force and competence through sport. Drinking, fighting, and adult sexual relationships predicated on male force and activity are examples of other bodily performances of 'hardness' through which young men signify a particular hegemonic working-class masculinity (Canaan 1996). Paul was no exception. He spent most of the time when he was not working prior to his accident drinking with his 'mates', playing football or dating 'girls' as he recalls in this quotation:

> We had a local neighbourhood pub that we used to go in. We used to meet. Then we'd go out round town, round pubs in town and that. That'd be on a regular thing. It depended on shifts, what shifts you were on, whether you went out. Whoever were on, and however many there was. That's about it. We started off in local. We used to meet up in the same one you see. Sometimes we stopped there all night, if weather were bad or whatever. If you're a bit hungry they used to do a pie and peas and things like that.

Writing about the general feminine style of bodily comportment, Iris Marion Young (1990: 146) argues that 'For many women as they move in sport, a space surrounds us in imagination that we are not free to move beyond; the space available to our movement is a constricted space.' And she goes on to claim that women lack self-confidence and underestimate their bodily abilities, stating that 'she often lives her body as a burden which must be dragged and prodded along and at the same time protected'. In contrast, as the discussion so far suggests, a hegemonic masculine style of bodily comportment is about having the freedom to move freely in space and to appropriate it both through physical displays of competence and force, and through having social confidence and a sense of personal security. As Connell writes:

> To be an adult male is distinctly to occupy space, to have a physical presence in the world. Walking down the street, I square my shoulders and covertly measure myself against other men. Walking past a group of punk youths late at night, I wonder if I look formidable enough.
>
> (1983: 19)

Prior to his accident then, Paul, despite the fact that his work and, to a certain extent, his leisure were physically demanding activities framed within masculine discourses of 'strength, sporting ability and sexual prowess', took his body for granted. He rarely reflected on the size, shape or appearance of his body, on his routine bodily practices and abilities, or on the extent to which his own narrative of the self was constructed through a particular shared and deeply embodied working-class masculinity, developed in the twin locations of the workplace and the pub. Leder (1990: 84) uses the term 'bodily disappearances' to describe the way bodies are characteristically taken for granted or overlooked in this way.

Doing disability: coming to terms with bodily impairment

Lifecourse studies have emphasised the importance of thinking about moments of 'transition' in people's lives, rather than the relatively fixed stages through which people are supposed to pass implied by the term lifecycle (Hareven 1978). An accident resulting in bodily impairment is obviously one such moment that can mark a dramatic re-alignment in an individual's way of living and identity formation – what Dyck (1995: 318) terms 'biographical disruption'. As Connell argues:

> Bodies cannot be understood as a neutral medium of social practice. Their materiality matters. They will do certain things and not others. Bodies are substantively in play in social practices such as sport, labour and sex.
>
> (1995: 58)

172

Paul's accident which left him paralysed from the chest down transformed his relationship with his own body. His injury immediately located him in a narrative of identity not of his own making – that of a 'disabled' person which in turn was framed within discourses of 'weakness', 'incompetence', 'dependence', 'illness' and 'impotency' (Butler 1998; Butler and Bowlby 1997) in contrast to the discourses and bodily performances of hegemonic masculinity, including, for example, strength and competence, within which his identity had previously been located. As Connell (1995: 54) has argued: 'The constitution of masculinity through bodily performance means that gender is vulnerable when the performance cannot be sustained – for instance, as a result of physical disability.' A position supported by Robert Murphy, an anthropologist who describing his own experience of being disabled writes:

> Paralytic disability constitutes emasculation of a more direct and total nature. For the male, the weakening and atrophy of the body threaten all the cultural values of masculinity: strength, activeness, speed, virility, stamina and fortitude.
>
> (1990: 94, cited in Gerschick and Miller 1995: 183)

From being oblivious to his body, in the first few years following his accident, Paul began to understand himself as a prisoner of his body because of the disabling socio-spatial environments he found himself in. Paterson (1998) argues that in everyday environments the impaired body 'dys-appears' both functionally and aesthetically in that it does not fit in, there is no place for it, yet through this very process of bodily dys-appearing, disabled people's attention is often drawn to their own bodies – reminding them of their impairment and that they are 'other'. The well documented (Imrie 1996; Kitchin 1998, 1999) difficulties that wheelchair users regularly encounter accessing and using so-called 'public' spaces, meant that Paul found it hard to get out and about independently following his spinal injury, especially until his compensation money came through and he was able to buy a car. Resentful of the need to plan ahead if he wanted to go out to the pub or a club, and frustrated by the patronising attitudes of shop assistants assuming that he was incompetent or unable to speak or to make decisions for himself, Paul withdrew from everyday public spaces.

The so-called 'feminised', 'privatised' location of the home, rather than the 'masculine', 'public' spaces of the workplace and the pub became the focus of his day and thus the most important sites in the construction of his own narrative of the self. In this way while Paul's accident had left him bodily impaired, his experiences of everyday socio-spatial environments dis-abled him. Foucault (1979: 96) describes power relations and acts of resistance as 'furrowing across individuals themselves, cutting them up, remoulding them'. Because Paul could no longer move his body in the same as before, he believed that he could no longer perform his identity in the same way either. He began to think of himself as dependent, and to rely on his mother to do most things for him, making little attempt to develop his bodily capacities and competence. Instead he spent his days eating out of boredom and watching television. His weight spiralled out of control, making it more difficult for him to lift himself and to push himself, further incapacitating or 'disabling' himself – both literally, and in terms of reinforcing the construction of his self-identity as 'weak', 'helpless' and 'dependent'. Paul described how his loss of physical ability was matched by a sense of relinquishing his masculinity and becoming a genderless – in his words – 'lump':

> So when I were at home all day, when you're not doing anything yer just sat about. If weather's bad and you can't get out and yer stuck in, I just watched telly all day and I'd go have a bag of crisps or a bar of chocolate what 'ave you . . . know when I first, when I 'ad me accident after that I did put a lot of weight on . . . when I got 'ome sort of thing 'cause I weren't doing nothing at all – it were just seem to be – be like going fishing, so there were no exercise in that . . . So I did put a lot of weight on then.
>
> I hated meself for ages cause I didn't think of myself as a man. 'Cause when, like when you'd been someone who'd been like strong and 'ard working its 'ard to be stuck in on your own, sat about like a lump.

Connell (1995: 61, paraphrased) has argued that bodies are both objects and agents of practice and that practices themselves form the structures within which bodies are appropriated and defined. He termed this 'body-reflexive practice'. Paul's experiences demonstrate the circuits involved. His bodily impairment had social consequences in that he was dis-abled by socio-spatial environments which led to him withdraw from public space, this in turn led him to gain weight and so his bodily disgust and discomfort further dis-abled him and undermined his gender identity. To paraphrase Connell (1995: 61) further, this is not just a case of social meanings being imposed on to Paul's body, rather his body-reflexive practice calls them into play, while his bodily experience facilitates the circuit.

Doing disabled masculinities: narratives of the self

On a trip to buy a new wheelchair Paul met Mike who had been in hospital at the same time as him. During their rehabilitation they had both played basketball. Mike now ran a basketball team and persuaded Paul to attend a training session. Paul rapidly became a good player for his level of spinal injury and was selected for the British national squad. As a result he constructed a new narrative of the self – that of international athlete – which contradicts the narrative of identity not of his own making – that of a disabled person – within which he is located, and which is framed within discourses of 'weakness', 'incompetence' and so on. In contrast to his footballing days when he took his body for granted, as a basketball player Paul has become very self-reflexive about his body. As part of the team's training a doctor was asked to give each player independent dietary advice. Paul abandoned his diet of pie and chips and began to eat pastas, vegetables, salads and fruit. He also began to work out in order to try and reduce his body weight and increase his strength and therefore both his speed and his capacity to lift himself. In this way he has developed his own mobility. Water rather than beer is now his main fluid intake. As a result of drinking water to avoid dehydration while playing basketball, he has also found that he has suffered less frequently from bladder infections and that his general health and consequently bodily capacities and sense of well being have improved. And he has begun to pay more attention to his body shape and to 'working on' different parts of his body. As a result, even though he is bodily impaired, Paul perceives himself to be more 'embodied' (i.e., aware of and in touch with his body) and 'able-bodied' (i.e., in terms of his physical health, diet and fitness) than before his accident. This is reflected in the following quotations from conversations with him:

> We 'ad somebody a few year ago before went on to this tournament an' all and for t' Europe Cup, for a while and 'e put us on this diet to try and lose a bit of weight. So 'e gived us all dietary things what to eat properly and when to eat, you n'aut to go out. You want to eat two hours before a game, for it to digest and for it to go into your system for energy sort of thing, you know, carbohydrate, an' a protein and what have you, so – an' as little fat as possible.
>
> I don't think before I 'ad me accident – I don't think I really thought – like I said I don't think I thought about it [his body], because when yer like eighteen year old you don't really think about it because you've got enough energy I think to do anything whatever size, whatever you are. I think I'd got enough energy to – and plus I weren't playing sport then at top like I am now, playing like I said for t' GB squad. So really it's – can't get no 'igher. And wi' football or whatever before that I weren't that bothered about it really. So and like I say I ain't done nothing for what six, seven year really. Nothing I ju . . . I do a bit of fishing what have you. Then till I took this up – an' it has changed me I think. It has changed me habits as well because I try not to put weight on. Because if I put weight on it's gonna make it harder to push around on court as well. But it's not just basketball it's everyday like, everyday

really, 'cause you've got to carry your weight about everyday – with lifting out and even lifting on to a dentist chair which is high up – you can have problems with that if you've got a lot of weight to lift up and anything like that.

Me legs are pretty big and I'll never lose that. But I've lost a bit off there but if you go on a diet sometimes to lose a bit of weight you lose it off yer shoulders and arms 'cause yer using them all't time. So you have to keep on t' weight to keep them up or otherwise you get too weak. In which my arms and shoulders are everything really, 'cause I 'ave to lift meself with them in an' out of bed and out o' the bath, car and everything so, so it's all that really.

Before I were prone to these infections – bladder infection sort of thing and they did throw you out – I were in bed like for two or three weeks with them. And you're temperature's up and then you're freezing cold, but you're shivering, you're body's all shivering but you're warm – you're sweating, it's, it's terrible. But since basketball I don't seem to have been affected with 'em.

Basketball has empowered him to resist negative discourses and narratives of disabled identities by creating a different subject position for himself. Through basketball Paul has re-defined some of the hegemonic characteristics of masculinity – developing his bodily competence and skills; and his sense of independence within the limitations of his impairment. Rather than limiting his mobility, para-doxically his disability has stretched his geographical horizons. As an able-bodied 'working-class' man he was firmly rooted in his home town, rarely travelling beyond the nearest city. As an international sportsman he has gained access to privileges previously denied him, travelling all over the UK, to Europe and to North America. As a result of constructing a narrative of the self through the national and global spaces of domestic and international basketball competition, Paul has in turn grown in self-confidence and begun to take up and appropriate space again – participating in a masculine sub-culture – this time through the basketball team (one which is still predicated on sport, banter and drink).

On the one hand then, Paul appears to be constituting his masculinity through a series of bodily performances that reject discourses that equate bodily impair-ment with weakness, incompetence and so on, and which rely instead on his ability to reproduce a range of bodily performances that are associated with hege-monic masculinities such as strength, skill, etc. On the other hand, his experiences of being disabled have also led him to reconstitute his gender identity in subtly different ways. For example, he now spends more time talking and listening to others, and while he once completely relied on his mother to cook for him, he does now take a more active role in the kitchen, often cooking his own pastas and special meals. In this way, his narrative of his own identity represents a complex blend of hegemonic masculine bodily performances and practices and new constructions of embodied masculine performances and practices that stem from his experience of being dis-abled.

Once again Paul's experiences demonstrate the circuits involved in Connell's notion of body-reflexive practice. The bodily pleasures he derived from playing basketball (in terms of sense of health, well-being and so on) had social effects in that he developed new friendships, changed his lifestyle and expanded his geographical horizons. These changes in turn resulted in him gaining a sense of self-confidence in his bodily performances and his ability to occupy and take up everyday space again. Thus the circuit goes in this case from bodily interaction and bodily experience to the creation of new social relations and a subtly reconstituted sense of identity based on these bodily performances.

Conclusion

The case study of Paul highlights some of the processes through which we understand ourselves as embodied or able bodied; how we can be located by discourses of gender or ability/disability within narratives not of our own making; and how individuals negotiate their own individual identities through different circuits of body-reflexive practice. It also highlights the interconnectedness of the body to other spatial locations, specifically the role of social relations in locations such as the workplace, the home or the pub in shaping individual narratives of the self.

Paul's reconstitution of his masculinity in the light of his spinal injury can be read critically then, as a 'triumph over adversity' story in which the 'hero' achieves success despite, or perhaps because of, his bodily impairment. While Paul's experience is not typical, and this is not an empowering analysis for those whose bodies or material circumstances prevent them from re-defining their lives in quite such drastic and high profile ways; at the same time it is important to recognise that others – who may be more restricted by their impairments – can also renegotiate or reconstitute their identities around their changing bodyspace (albeit within different enabling or constraining circumstances). Isabel Dyck's (1995) paper about the changing lifeworlds of women with multiple sclerosis includes two case study examples of women who have renegotiated their everyday spaces very differently in the light of a shared chronic illness. Indeed, as a range of recent geographical writing on the body in relation to ageing, childhood, pregnancy and so on has demonstrated – for example, Robyn Longhurst's (1994) study of pregnant embodiment – to greater or lesser degrees we are all actively engaged in reconstituting our self-identities through bodily reflexive practices in relation to the way our bodies are discursively constituted.

Paul's story is perhaps a story that should not be too readily dismissed, helping to counter as it does what Michael Dorn (1998: 198) has described as the tendency of 'All too many programs and services for the disabled [to] *presume* passivity on the part of the consumer – a blank slate upon which institutional protocols are inscribed.' Drawing on the experiences of disabled political activist Patty Hayes, Dorn (1998: 196) explores how one woman's challenges to the dis-abling spatial barriers within her everyday environment 'resonate across wider domains effecting broader societal changes'. In contrast to Patty Hayes, who has taken her 'bodily struggles for access and mobility out of her house and into the streets, with the goal of changing her neighbourhood' (Dorn 1998: 195), Paul himself does not see the problems he has experienced coming to terms with his spinal injury as lying with dominant discourses of masculinity, disability and with dis-abling socio-spatial environments. Rather than being politicised by his experiences and using them as a platform from which to campaign against the ableist assumptions of his everyday environments, Paul has renegotiated a complex individual identity as an international athlete (which partially derives from hegemonic working-class masculinities while also reworking this identity in subtly different ways) which has allowed him to expand his own lifeworld. While, his actions do not challenge the ableism of the urban fabric in the same way that Patty's campaigning does, exploring the processes through which individuals like Paul re-negotiate their identities can help to expose the inadequacy of contemporary Western society's conceptions of what it means to be disabled. In this way, the experiences of unpoliticised individuals, such as Paul, may also contribute to informing broader political projects that are dedicated to challenging the complacency of ableist society.

Acknowledgements

I am grateful to Ruth Butler, Janet Elsden and Peter Jackson for lending me key books and articles from their personal 'libraries'. I acknowledge the support of the Leverhulme Trust which funded the research on which this chapter is based (award number F118AA). Finally, many thanks to Ruth Butler and Hester Parr for waiting patiently for me to deliver the goods and for their helpful editorial comments.

References

Bell, D. and Valentine, G. (1997) *Consuming Geographies: You Are Where You Eat*, London: Routledge.

Butler, J. (1990) *Gender Trouble*, London: Routledge.

Butler, R. (1998) 'Rehabilitating the images of disabled youths', in T. Skelton and G. Valentine (eds) *Cool Places: Geographies of Youth Cultures*, London: Routledge.

Butler, R. and Bowlby, S. (1997) 'Bodies and spaces: an exploration of disabled people's use of public space', *Environment and Planning D: Society and Space* 15, 4: 411–33.

Canaan, J. E. (1996) '"One thing leads to another": drinking, fighting and working class masculinities', in Mac An Ghaill, M. (ed.) *Understanding Masculinities: Social Relations and Cultural Arenas*, Buckingham: Open University Press.

Connell, R. W. (1983) *Which Way is Up? Essays on Sex, Class and Culture*, Sydney: Allen & Unwin.

—— (1995) *Masculinities*, Cambridge: Polity.

Dorn, M. (1998) 'Beyond nomadism: the travel narratives of a "cripple"', in N.J. Nast and S. Pile (eds) *Places Through the Body*, London: Routledge.

Dorn, M. and Laws, G. (1994) 'Social theory, body politics and medical geography: extending Kearn's invitation', *Professional Geographer* 46: 106–16.

Dyck, I. (1992) 'Health and health care experiences of the immigrant woman: questions of culture, context and gender', in M. Hayes, L. Foster and H. Foster (eds) *Community, Environment and Health: Geographic Perspectives*, Victoria: University of Victoria Press.

—— (1995) 'Hidden geographies: the changing lifeworlds of women with multiple sclerosis', *Social Science Medicine* 40, 3: 307–20.

Foucault, M. (1979) *History of Sexuality, Vol. 1 An Introduction*, London: Penguin.

Gerschick, T. J. and Miller, A. S. (1995) 'Coming to terms: masculinity and physical disability', in D. Sabo and D. F. Gordon (eds) *Men's Health and Illness*, London: Sage.

Hareven, T. (ed.) (1978) *Transitions: The Family and The Lifecourse in Historical Retrospect*, New York: Academic Press.

Imrie, R. (1996) *Disability and the City: International Perspectives*, New York: St Martin's Press.

Kearns, R. (1993) 'Place and health: towards a reformed medical geography', *Professional Geographer* 45: 139–49.

Kitchin, R. (1998) '"Out of place", "knowing one's place": space, power and the exclusion of disabled people', *Disability and Society* 13, 3: 343–56.

—— (1999) 'Creating an awareness of Others: highlighting the role of space and place', *Geography* 84, 1: 45–54.

Leder, D. (1990) *The Absent Body*, Chicago: Chicago University Press.

Longhurst, R. (1994) 'The geography closest in – the body . . . the politics of pregnability', *Australian Geographical Studies* 32: 214–23.

Milburn, K. (1995) 'Never mind the quantity, investigate the depth', *British Food Journal* 97: 36–8.

Mitchell, J. C. (1983) 'Case and situation analysis', *Sociological Review* 31, 2: 187–211.

Morris, J. (1991) *Pride Against Prejudice: Transforming Attitudes to Disability*, London: Women's Press.

Murphy, R. F. (1990) *The Body Silent*, New York: Norton.

Oliver, M. (1990) *The Politics of Disablement*, Basingstoke: Macmillan.

—— (1996) 'A sociology of disability or a disablist Sociology?', in L. Barton (ed.) *Disability and Society: Emerging Issues and Insights*, London: Longman.

O'Neill, T. and Hird, M. J. (1998) 'Double damnation: Gay disabled men and the negotiation of masculinity'. Paper presented at the British Sociological Association, Edinburgh, April. Available from: Myra Hird, Dept. of Sociology, University of Auckland, Private Bag 92019, Auckland, New Zealand.

Paterson, K. (1998) 'Disability studies and phenomenology: finding a space for both the carnal and the political'. Paper presented at the British Sociological Association, Edinburgh, April. Available from: K. Paterson, Glasgow Caledonian University, Glasgow.

Platt, J. (1992) 'Case study in American methodological thought', *Current Sociology* 40, 1: 17–48.

Valentine, G. (1998) 'Food and the production of the civilised street', in N. Fyfe (ed.) *Images of the Street*, London: Routledge, 192–204.

—— (1999) 'Imagined geographies: geographical knowledges of self and other in everyday life', in D. Massey, J. Allen and P. Sarre (eds) *Human Geography Today*, Cambridge: Polity Press, 47–61.

Valentine, G. and Longstaff, B. (1998) 'Doing porridge: food and social relations in a male prison', *Journal of Material Culture* 3, 2: 131–52.

van Velsen, M. (1967) 'The extended case method and situational analysis', in A. L. Epstein (ed.) *The Craft of Social Anthropology*, London: Tavistock.

Yin, R. K. (1984) *Case Study Research: Design and Methods*, London: Sage.

Young, I. M. (1990) *Throwing Like a Girl and Other Essays in Feminist Philosophy*, Bloomington: Indiana University Press.

10

BODIES AND PSYCHIATRIC MEDICINE: INTERPRETING DIFFERENT GEOGRAPHIES OF MENTAL HEALTH

Hester Parr

Introduction: towards new geographies of mental health

[I]t is now possible to identify a small field of geographical studies tackling various aspects of how space, place, environment and land-scape impact upon people with mental health problems. A simple distinction here, albeit not a rigid one, can be drawn between these studies concerned with *geographies of mental ill-health* and those con-cerned with *geographies of mental health facilities*: the former concen-trating on links between sufferers of mental ill-health and where they live, work and move, and the latter concentrating on ramifications of where facilities designed to treat sufferers are physically located.

(Philo 1997: 73)

Within the discipline of geography, mental illness and mental health have long been phenomena which have prompted both debate and a range of theoretical approaches and empirical studies. Philo's comments here are useful, and help us to distinguish between two trajectories of academic inquiry in the geograph-ical literature: mental illness studies and mental healthcare studies. Although Philo does not here problematise the language employed by geographers which arguably reinforces notions of 'suffering' and 'illness', he nevertheless draws attention to the different ways in which phenomena conceived of in terms of mental health and mental illness have been *spatially* investigated. This chapter will document some elements of a research project (Parr 1997a) which expand upon both of the identified trajectories of inquiry (the foci upon both 'illness' and its treatment), but in ways which emphasise different spatial considerations

181

to those conventionally considered in geographies of mental health. More specifically, this chapter considers the spatiality of the body as a site through which people with mental health problems experience, conceptualise and resist mental health problems and subsequent medical treatments.

It should be noted at this early stage that many geographers have not thoroughly examined what is meant by the term 'mental illness', with much work being uncritically focused around particular medical diagnoses such as schizophrenia (Giggs 1973) and depression (Dean 1979). Although this issue is not addressed at length here, the term 'mental illness' and various diagnoses associated with this pathologisation are not accepted as the final word on how alternative mental states can be understood. These different states (embracing both mind and body) often entail people thinking, feeling, behaving, moving and speaking in ways which are considered different to common social norms (see Parr 1997b). It may be useful to regard these differences as a variety of 'madness': a term which does not invite any essentialised or medical explanation, but rather refers to mind/body differences which are individually distinct. Many people do feel themselves to be 'ill' when they experience these differences, although naming particular mental states as 'mental illness' (and subsequently administering medical treatment within an institutional geography of hospitals and clinics) may be only *one* (quite recent) interpretation.

Here Philo can be useful to us again:

> [In] Foucault's classical age – the forces of reason (as they become crystallised and dispersed across a range of intellectual and practical realms) gradually identified, pathologised and intervened in the elements of unreason (as a broad sweep of passions, attitudes and behaviours manifested in a multiplicity of ways).
>
> (1996: 4)

The idea that unreason (or madness) has been increasingly separated from reason is a Foucauldian notion which identifies historically the processes through which a medicalisation of madness into mental illness has occurred, itself bound up with relationships of power and knowledge. Geographers have been slow to recognise these relationships, and have often simply investigated the spatial patterns of diagnosed mental illness or traced the spatial distributions of ex-mental patients. We should perhaps recognise that there may be other interpretations (and other geographies) of these mind/body differences; particularly ones which emerge from the individual concerned. It is perhaps important that a geography of mental health takes into account these 'alternative' spatialities, ones that are also sometimes divorced from a medical–institutional geography. Here I am referring to many possibilities from the more 'structural geographies' of political-economy as they impact upon people, treatments and institutions (Scull 1977; Warner 1985), to more 'agency-orientated geographies' of embodied everyday lives, experiences, identities and imaginings (broadly where this chapter can be situated). With reference to both of these orientations, but particularly the latter, the geography of the body has been neglected when considering mental health, and it is the purpose of this chapter to consider the possibilities of such a focus of concern.

It is important to acknowledge that many individuals *do* accept medical diagnoses for their experiences of different mental states, and many find that psychiatric medication can alleviate stressful or confused states-of-mind. However, some individuals, not least those interviewed as part of the empirical work featured here, sometimes resist a totalising medical 'naming' of their states of mind/body. Hence, when we discuss a 'geography of mental illness and mental health', we should be doing so critically, with an eye to the alternative definitions and understandings that individuals and groups have of their own mental states. This questioning of terminology can be seen in some ways as a questioning of the medical authority that particular diagnoses appear to hold, and of the subsequent social power (and powerlessness) that such a naming can hold for an individual. Groups or individuals may even want to appropriate their own names and categories for the experiences which they have, and the examples used below include the terms 'people with mental health problems' and 'users' (of psychiatric services) as these were phrases routinely employed by the individuals who informed this particular account. Although they perhaps still hold medical connotations, both are meant as politically different from terms such as 'the mentally ill'.[1]

One possible route for studying 'alternative' spatialities of mental health, taking into account the concerns raised above, may be through a consideration of embodied experience. Although perhaps a crude point, it is important to remember that the person with mental health problems is a body as well as a mind and that different states of being (or 'problems') effectively bleed between the mind and body, a key point made within the introduction to this book. Turning, then, from a critique of mental health studies and terminology, this chapter will briefly introduce some relevant debates which have surrounded geography and the body. Following this, an empirical case study will show how concerns about the body and geography might inform writings on geographies of mental health, ones which build upon, but also differ from, the two conceptual divides which Philo identifies.

Bodies and geography

Regarding the body as a site both symbolic and real is something which geographers have recently begun to experiment with and write through. From feminist critiques of surface inscriptions and cultural meanings which are constructed, conveyed and experienced bodily to the acknowledgement of corporeal differences, ableist norms and an embodied politics of resistance to such norms, geography is opening itself up to the body (Bell *et al.* 1994; Bell and Valentine 1995; Butler and Bowlby 1997; Cream 1994; Dorn and Laws 1994; Duncan 1996; Longhurst 1994, 1995; McDowell and Court 1994; Nash 1996; Parr 1998; Rodaway 1994; Rose 1993). In terms of geographical writings which are explicitly concerned with disability and health, Dorn and Laws (1994) have indicated the ironic lack of attention towards bodies, especially in medical geography. By drawing on social theory, they make the distinction between the phenomenologically 'lived' body (following the work of Merleau-Ponty 1962) and the body as a constantly reworked surface of inscription (following the work of Grosz 1993); both of which could be useful in medical geographical analysis. Especially with regard to the latter categorisation, the body and its many differences are increasingly being recognised by geographers. Some regard it as a site of oppression upon which social norms and inscriptions etch themselves in potentially stigmatising and exclusionary ways (Sibley 1995), but others discuss it as a site of resistance wherein emancipatory body politics both reflect and continue everyday social and material struggles (Dyck 1995). More specifically, Dorn and Laws argue for geographically sensitive body politics: 'geographies of the deviant body, something we see as a central focus of a "reformed" medical geography, need to balance the subjective and objective experiences of place by contrasting "insider" and "outsider" views of emplacement' (1994: 108). By drawing upon these theoretical influences it is thought that empirical work can take account of the politics of mentally different bodies (not a contradictory term if we understand the mind and body as part of a continuum and not as a dualism) as sites which are socially and spatially excluded (a key concern of geographies of mental health), but also as sites where imaginations, feelings, emotions and distress reside (indicative of aspects of self and identity: see Parr and Philo 1995). As Pile notes: 'If people's actions are to be understood, then it is necessary to understand that feelings, impulses and thoughts are somewhere in the flesh' (1996: 87). The implications of such a statement are many, but it is hoped that this chapter will show that mentally different bodysites are not only complex spaces of medical inscription and routinisation but also of self, identity, emotion, relief and resistance. This focus is informed by thinking which challenges traditional Western rationalism, which forces apart considerations of the mind and body, and serves to open up the discipline of geography to a diversity of subjective and messy/bloody/seeping/orgasmic/painful knowledges (ones previously ignored as sources of academic understanding). This deliberate emphasis upon real bodily experience, and on associated contestations of social inscriptions prompts themes which surface in the following case study materials. The emphasis on the intersection of body and mind in the context of the everyday experiences of psychiatric service users is a reflection of the very real politics of psychiatric medication use and also a reflection of concerns within geography and the wider social sciences to abandon artificial divisions between mind and body in conceptual writings.

Psychiatric services users in Nottingham

Nottingham Advocacy Group (NAG) and Nottingham Patient Council Support Group (NPCSG) are two parts of one organisation comprised of ex- and current users of psychiatric services in Nottingham, UK which has existed since 1986. The organisation was formed with the explicit intention of critiquing medical–psychiatric environments and working practices.[2] These critiques and the subsequent 'political' actions have been focused around a variety of tasks and spaces, from pressing ward managers to replace bath plugs, to providing legal support to individuals sectioned under the British Mental Health Act, to meeting Healthcare Trust managers to discuss the planning of future mental health services. The NPCSG in particular comprises a network of individual patient councils across the city's medical–psychiatric sites, all run and supported by key user-volunteers. The medical–institutional geographies which 'the users' often critique and work within are many and varied, occupying different scales from the ward to the community mental health centre to the city. Some very specific aspects of these geographies will be discussed in this chapter (for other aspects of this research project and comments upon how these scales interlink see Parr 1997b), and attention will be paid to how the user organisation has begun to politicise the body when critiquing uses of psychiatric medication. This theme, although selective, is useful in thinking through how people who are labelled as 'mentally ill' articulate their individual and collective human agency through different spaces to those previously identified in geographical studies concerning mental illness and mental health care.

The user organisation and medical critiques

A decade ago, when patient advocacy was just beginning in Nottingham, the main aims of the NPCSG were (and have continued to be) issue-based, and have involved patients voicing opinions about their everyday institutional environ-ments. In early literature produced by the group, indications are given about the concerns generated by the first few meetings:

Issues raised	
Living conditions	Layout of ward/day centre
	Toilets/shortage of toilets
	Access/lack of access
	Privacy/lack of privacy
Activities	Daily activities/lack of activities
Catering	Rations
Information	General
	Specific: medication/lack of general
	information/lack of specific legal rights information
Treatment	Purpose of treatment/means of treatment
	Information about treatment

Source: NPCSG 1986: 28

The range of issues indicates the breadth of concern about the everyday functioning of care environments in the mid-1980s, and the potential of the group
to challenge medical and psychiatric hegemony must have seemed considerable
at the time. Replacing toilet paper was an issue for the emerging councils, but
then so was information about treatments and new demands for explanations
from medical staff. In the emerging national debates about community care,
coupled with patients becoming organised in order to critique working practices
and therapeutic regimes, the changes must have seemed challenging to some
mental health workers. Academic debates which covered issues of social policy
around this time were striking in their consideration of empowerment issues,
deconstruction of the medical hegemony of psychiatry, and critiques of the 1983
UK Mental Health Act (Rogers and Pilgrim 1989; Tancredi 1983; Turner 1987).
The anti-psychiatry movement, which had begun life in the 1960s with the arguments of commentators such as Szasz (1961) and Scheff (1966) had not abated,
and if anything had been revived by the progression to community care policies
during the 1980s. Administrative and medical personnel involved in the provision of psychiatry services would have been aware of these interrelated academic
and patient critiques increasingly impacting upon the practice of mental health
care. It is out of this context that users of psychiatric services began to piece
together a powerful, if erratic, critique of psychiatric medicine in Nottingham.

Mind/body space, medication and resistance

The Crusader representative asked if it was just a question of economics
that chemotherapy was still the main treatment for people with mental

health problems and why more resources were not given over to psychotherapy and talking treatments? . . . Mrs Smith for the Health Authority explained the differences between primary and secondary care and indicated that 9 out of 10 people with mental health problems do not come into contact with mental health services and were dealt with by general practitioners, and that counselling services may be more appropriately incorporated into a GP practice.

(Citywide Council Minutes 24 May 1993: 2)

I think there is an over-emphasis on medication and an under-emphasis on talking treatments, and in addition to that I think there is an under-emphasis on helping people get back on their feet in new settings. I think that, when people have mental health problems, they need not so much to be medicated, but to clarify where their problems have come from and then to help them escape from the source of their problems . . . What tends to happen is that people are seen by a psychiatrist, given a medical diagnosis and pumped full of drugs and get left very much to their own devices on a ward.

(Interview with NAG worker 14 July 1993)

It is arguably the case that the treatment of psychiatric patients through the administering of medication is one of the most powerful social acts that occurs within medical places (Illich 1977) incorporating many political dimensions, as the quotations above indicate. The introduction of psychotropic drugs on a large scale at the end of the 1950s has boomed into a multi-million pound pharmaceutical industry, and has been hailed as one reason behind the trend of deinstitutionalisation. The mystique and scientific knowledge surrounding the introduction of powerful psychiatric drugs has prompted critiques of this medical intervention as a form of social control:

Since sickness can be regarded as a form of social deviance, the medical profession has a policing function within society. Medicine as a form of social control involves the standardisation of illness into phenomena which can be managed by bureaucratic agencies.

(Turner 1987: 214–15)

Such concerns about psychiatric medicine have not only been applicable since the 1950s. Many historical studies have emphasised the complex magical, medical, moral and social contexts to different treatment regimes and therapeutic spaces (Philo 1992; Porter 1989; Scull 1977). Treatments for madness and psychiatric conditions have undoubtedly varied in time and space, but as Jones (1983)[3] notes, can arguably be seen to have taken three broad directions: those treatments concerning the soul, mind and spirit, those concerning the physical body, and those concerning the built or natural environment (although these 'fields of intervention' have mixed confusingly with each other in particular circumstances). In academic geography, attention has already been paid to the third dimension, that of built or natural environments (Philo 1992; Smith 1980), although these sorts of writings inevitably draw upon the two other dimensions of Jones' typology, especially when considering the medical and moral dimensions of nineteenth-century institutional and twentieth-century deinstitutional psychiatric geographies. Throughout human history the search for causes and treatments of mental health problems has involved a multitude of everyday materials and places from 'temple sleep' (Jones 1983) to the healthy aspects of slopes and vegetation (Philo 1992), to hydrotherapy, bloodletting and early medicinal treatments with herbs (Jones 1983), to asylums (Philo 1992), to community care (Kearns 1990). The thinking behind such different treatments of mental distress have been as varied as their methods (despite Turner's understanding of an underlying search for social control). For the purposes of this chapter, however, I am most interested in references to the somatic, the physical bodily treatments, and particularly the uses of water, herbs and temperature with respect to the body. The thinking surrounding such corporeal treatments could be characterised as representative of a mind–body dualism in which the mind was viewed as subservient to the body, and hence as manipulable through it (arguably a reversal of more 'usual' hierarchies of dominance in reference to the mind–body). This view still pervades modern psychiatry, where powerful chemicals are administered and absorbed by the body in order, supposedly, to control (or even to cure) the symptoms of a 'mind diseased'. In terms of mental health, then, the body has had extraordinary significance as a site for which treatments have been prescribed and on which certain understandings of both illness and 'deviance' have been inscribed.

Interestingly, the manipulation of 'nature' and 'natural ingredients' in relation to the body, and for the direct relief of 'unhealthy' mental symptoms, has a long history. Since classical and medieval times 'the humours' ('black bile' being associated with mania and 'yellow bile' being associated with melancholy: Philo 1992: 42) have been seen as amenable when treated with a varied arsenal of herbal 'cures': for example hellebore for the relief of melancholy, madwort for abating rage, and with bad humours being vanquished with periwinkle and rosemary (Jones 1983: 36). Such seemingly antiquated notions of herbalism (as opposed to contemporary chemically complex, mass manufactured scientific medicines) have been remobilised (Gesler 1992), both in general, as a feature of (what might be termed) late twentieth-century capitalist neurosis (see Plate 10.1), and in particular, as part of a resistance by the users of Nottingham mental health services. Such resistance, as will be documented, entails a potential destabilising of the perceived authority of psy-

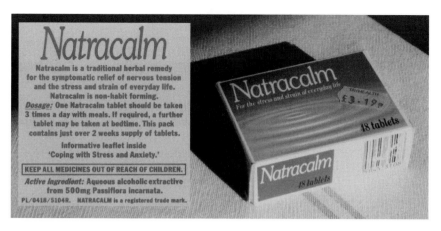

Plate 10.1 Herbal medicines for mental health[4]
Source: Author

chiatric medication in favour of 'alternative', or in this case, 'complementary' medicines and 'talking treatments' (representative of more psychoanalytic approaches[5]). In arguing for a choice and combination of such 'treatments', the users of Nottingham can be seen as requesting therapeutic interventions which serve to link both the mind and body, while also destabilising the authoritative discourse of conventional psychiatric medicine. The mind/body as a complex site of ill or mad experience, therapeutic intervention and resistance is the geography at stake here.

Experiences of the medicated body

Whether madness or mental health problems amount to a state-of-being caused by organic abnormalities, changes in body chemistry, social stress or is better regarded as encompassing extremes of human personality, the effects of such mental states are devastating to some individuals:

> I wanted to stay in bed for hours on end and I didn't want to come out. It's like your head is in a black cloud, really awful. You can't lead a normal life.
>
> (Interview with young woman 12 May 1995)

For many people who experience mental health problems the disruptions to both their senses of self and their everyday routines, including their feeling of control over time and space, are extremely distressing realities which prompt varied strategies of coping. For some, the ability to regain control and routine in time and space is intimately tied to the effects of the powerful medications that help control 'symptoms':

I'm on quite good medication which has helped me normalise on lots of things really, the combination that I take seems to suit me, there are quirks and different sorts of side effects that can affect you at different times of the day . . . like feeling dozy in the evenings or really knocked out in the mornings . . . that's what I usually have. It's an awful job to get up in the mornings, not because the tablets have knocked you out, but because they have worn off. The first part of the day is hell until you can get hold of your first cup of tea and a tablet. Then you can wash and face the day. It's all part of the routine really and I find that a huge part of my routine is really adjusting to the fact that the medication doesn't last all night.

(Interview with middle-aged male 29 June 1993)

Despite the positive tone of the above quotation, for many people interviewed in Nottingham the regulation of mental states by medication can lead to a perceived lack of individual control over the body, and there is undoubtedly a sense in which people's socio-spatial lives may be dictated by medical treatment as individuals taking medication cope with, and try to regularise, its effects. Many users would not dismiss the usefulness of certain forms of psychiatric medication in helping them to overcome some of their distress, but the reality of being prescribed such powerful drugs often leaves bodies feeling out of control. Severe side effects from medication can result in visible signs that may further serve to stigmatise the individual (through bodily actions which mark the individual as 'different'):

There's something strange that can be noticed about medication, for the simple reason it does things to your nervous system, not just tremors, but you can have a Stelazine kick, your foot can be going all the time.

(Interview with middle-aged male 29 June 1993)

Some resist such prescriptions, symbolically and literally, and it is common (especially amongst young and homeless people) that non-prescription drugs are used instead to relieve unpleasant mental states.[6] However, for these young people, the psychiatric services also hold opportunities for other desired medications to be obtained:

There's a Dr up there [Mapperley Hospital], a psychiatrist – he always gives you valium. You can get anything you want out of them – lithium, valium [lists others]. I've seen psychiatrists and I just do their head in, they can never work me out, I just tell 'em what I want and I get it, we have a bit of a laugh really.

(Interview with young male in homeless hostel: research diary
13 April 1994)

Such attitudes to medication are not common amongst older and homed residents. Individuals would often represent their different states-of-being as ones which need regulation by medication, and accepting their medication in this respect is often seen as part of being (or becoming) a responsible citizen. In some ways, to accept medical treatment means becoming acceptable to social and legal norms that govern behaviour in local communities:

> If you didn't take medication, you wouldn't be washing, you wouldn't be having somewhere to go during the day, you would really turn day into night and be making a fool of yourself round the streets round here, because you were so disorientated you couldn't make a decision about your self.
>
> (Interview with middle-aged male 19 April 1994)

Investigating alternatives in mind/body space: beyond medication

> They need a diagnosis in order to put you in a certain category of drugs, they are not really asking you questions that will lead on to some form of help by going to counselling or anything like that . . . The psychiatric profession today . . . they have evolved into drug pushers and we are just junkies . . . we have similarities with junkies in that we don't really know the dangers of what we are taking when we start taking them, and these drug pushers get us hooked.
>
> (Interview with NPCSG volunteer 29 September 1993)

A key aspect of the Nottingham user movement's activities has been to question the uses of medication in the treatment of people with mental health problems, and to critique the different experiences briefly characterised in the above section. As part of such a critique some users have begun a search for radical alternatives, and NAG and NPCSG have considered the potential of alternative and complementary medicine. NAG in particular has taken up this challenge of looking beyond psychiatric drugs for different ways of helping people with mental health problems, at one point holding a series of workshops that brought together practitioners of complementary medicine with mental health workers and users. NAG, and in particular the organisation's development worker,[7] has researched different forms of complementary medicine such as aromotherapy, reflexology, acupuncture, homeopathy and body structure therapies.[8] The group is categorical in stating that they have not been conducting a 'scientific study', but rather have looked to explore some ideas that may eventually have benefit to some users. This process is still being carried out over a long period of time, initially independent from the statutory authorities. These explorations can be seen as a politics of the body/mind as connected spaces or sites of health care.

To consider some aspects of the mixed conceptual/practical lens through which the users have begun to consider these different (if unusual) options, it is appropriate to reiterate the view that the body can be seen as a site upon which are projected various explanations/inscriptions of the self. This understanding is one which surfaced when, for example, NAG considered the benefits of massage for people in deeply stressful situations. Here particular geographical and contextual concerns about the body became evident. The users acknowledged in written reports that the locations (in physical and body space) where therapies take place are of crucial importance, and are linked to the personal and the emotional: 'Clearly a difficulty is where a patient is so seriously disturbed that the touch involved in these therapies is too emotionally explosive. In this respect reflexology is probably the least threatening form of touch. Much thinking and care needs to be given to the contexts, the circumstances and settings' (Davey 1993: 2).[9] The emphasis here on the body as a site is interesting, and is testament to the wider politics of negotiating a basis for thinking through mind and body connections without the chemical interventions of traditional medical discourse. This may sound a grand project, and yet other aspects of the work of the user movement are also about finding ways to legitimate alternative therapeutic interventions in conducive locations. This has taken the form of creating beneficial projects for users to be involved in locally (such as 'Ecoworks', a mental health gardening project),[10] as well as taking a longer term view concerned with challenging the metanarratives of psychiatric medicine.

In this sense, the harnessing of 'nature' and of 'natural ingredients' (evident in both the Ecoworks project and the complementary medicines research: see below) as beneficial to mental health can be seen as a return to past practices, beliefs and spaces. Importantly, though, this should not be seen as representative of old dualistic treatment strategies with the mind left subservient to the body, as these critiques of medication emerge from and exist within the wider context of the work of the users. For many years now NAG and NPCSG have also campaigned for more 'talking treatments', ways of managing mental health problems that do not rely so heavily on bio-chemical solutions and medical models of people's behaviour patterns. Users in Nottingham have now gone further to suggest that there are other ways of coping with mental health problems, ones which rely on *both* mind (more psychological or psychoanalytic approaches) and body (complementary medicines) therapies. Their personal experiences tell of such different approaches to treating mental health problems, ones which are *arguably* less devastating to senses of self than are the sole use of conventional psychiatric medication:

In January 1993, I had my last injection and two months later withdrawal symptoms erupted. After coping with insomnia, overactive mind and return of past symptoms, I contacted [an occupational therapist] to discuss practical advice. Lavender oil on my pillow helped me regain a natural sleep pattern; lavender oil in a daily bath, camomile tea helped calm me . . . complementary medicine working in harmony for me regained my natural weight and myself, the user, choosing with full knowledge of my psychiatrist . . . As my true personality has emerged from the suppression of the neuroleptic drug, the emotions of anger and frustration were eased by complementary therapy.

(Alexander 1993: 3)

As a result of reports produced from some NAG members, other users were able to receive aromotherapy in the district general hospital from a recently qualified health worker who had been involved in the NAG workshops. Her role here was to provide both counselling and complementary therapies to patients who wanted her services. The written reports were part of efforts by the user group to stress the importance of such developments, and hopefully to have complementary therapies made standard in psychiatric wards. This could be interpreted as a politics of the mind-and-body, which (in association with other therapeutic spaces) explores ways of contesting conventional psychiatry. In effect the user movement has sought to politicise the body as a site of connection between physicality and mental health by explicitly linking physical touch and psychological well being: 'many people at NAG and in the workshops spoke of relaxation and sleeping well after reflexology. This lends itself to the idea already mentioned that the various forms of touch therapy might be gentler forms of calming people and alternative to medication in some circumstances' (Davey 1993: 4). However, concerns about these alternative strategies have also been voiced in the promotion of these therapies to the local Healthcare Trust, and concerns include the potential for therapeutic abuse with touch therapies, particularly given that many users of mental health services have had experiences of physical and sexual abuse. The use of touch in these sorts of power-laden relationships can be problematic. The concern generated over physicality does not stop there, and the development worker for NAG makes explicit the politics of the body in the therapeutic relationship when discussing an example of how *he* was touched in a condescending way: 'Many psychiatrists seem totally blind to how power actually manifests itself like this in the interpersonal and emotional terrain, which is the real and proper link between the body and mental health' (Davey 1993: 5).

Davey uses older arguments from Lowen (1975) to help articulate his ideas about making mind and body connections. These links construct the body as a simultaneously real and symbolic site upon and within which mental health problems are manifest:

a fragmentation of the integrity of the personality [is] typical of a schiz-ophrenic personality. Schizoid means split. If the splitting exists in the personality, it must also exist in the body on an energetic level. The person is his [*sic*] body.

(Lowen 1975: 80)

Although these arguments are obviously situated within a particular context, and Lowen is arguing from a particular perspective, it is useful to think around these ideas of body and mind connections. Davey is perhaps exploiting the obvious links that he sees here in order to advocate holistic treatment of the mind and body as one, and not as torn apart in an artificial separation or intervention. He maintains that considerations (and manipulations) of the body have their uses in mental health: 'as people learn to use their bodies in different ways, releasing the holding patterns in muscles, forgotten and repressed memories and emotions can surface' (Davey 1993: 7). Moreover, Davey applies these thoughts about the body to critiques of psychiatry:

Critiques of the medical model are usually against the excessive and reductionist concentration of medical psychiatry on brain chemistry and the central nervous system, but there is a very valid connection between mental health and the physical body. Emotional blockage takes place in the muscular and movement systems of the body. These are just the systems damaged by psychiatric drugs.

(Davey 1993: 8)

Here the writings of the Nottingham users are cleverly addressing central tenants of medical psychiatry by reinscribing the body as a reservoir of the self (and hence inadvertently drawing on psychoanalytic perspectives: see Pile 1996), and then by emphasising how chemical interventions can damage, not cure, different body/mind states. These debates and thoughts about the body and mental health which emerge from the user movement in Nottingham are mixed-up and often contradictory, but none the less are powerful (and surprisingly conceptual) critiques of the geographies of the mind/body.

The uses of medication and related questioning of power relations are issues taken up on an *ad hoc* basis by the user movement in Nottingham, but they are nevertheless ones which encompass a variety of sites and scales from the body to the ward to the institutional geographies of the city and beyond. Direct challenges to the uses of bio-chemical treatments in the mental health services are few, even by this group, and – despite their investigations into complementary therapies – most psychiatric patients are not offered this service. The Citywide Council meeting, which has been a critical forum for negotiating many aspects of Nottingham's mental health service provision between users and 'providers', has seen little rigorous discussion of treatments, although some challenges have been documented:

> People expressed concern about not being taken seriously when looking
> at issues of treatment and in particular the administration of ECT.
> (Citywide Council meeting minutes 24 May 1994)[11]

Despite these public challenges in public forums, much more explicit are the personal accounts of disempowerment through medication and these were often related within interview situations, and rarely articulated within more public settings of the psychiatric sites. The practicalities of contesting psychiatric medication are immensely complex and difficult, and it is to this notion that this chapter briefly turns.

Investigating alternatives in institutional space: mundane practicalities

Despite what seems like a radical politics of the mind/body, the users in Nottingham have found it difficult to maintain a consistent critique of psychiatric medication. The erratic engagement with these issues is perhaps partly due to the varied demands on a small voluntary organisation, the personal difficulties involved in working when taking psychiatric medication, and the powerful discourses and medical personnel that surround such drug treatments. The users have found it easier to find more practical ways of liberating the body within medical regimes. Within the user's projects, the body as a site of potential emancipation is more linked to everyday life on the wards (which is evident in the table of issues at the beginning of this chapter), in which case the regulations of institutional care upon bodies in medicalised environments are challenged by the user movement:

> Occasionally we get complaints about patients. I remember one where
> they wanted to know why the nurses couldn't take this patient who
> never washed, who never changes their clothes, and dump her in the
> bath . . . [We argued] They [the patients] ought to be able to live the
> life they want to lead, they should be allowed not to wash and not to
> change, or wash their clothes.
> (NPCSG volunteer speaking at Wales MIND conference
> 27 November 1994)

> The patients at Mapperley still don't know that they can ask for food
> at night, including tinned food.
> (NPCSG meeting minutes 9 July 1993)

The rights of in-patients in a psychiatric ward to ask for food at night is important not only so that individuals do not go hungry, but also to reinforce
independence. To ask for food that is not at set ward meal times is a means of
displaying individual agency in a place where that agency is traditionally diminished. To exert the right for food, or to remain dirty, is to exert some control
over the body and can be interpreted as an expression of self (and not just as
an expression of a psychiatric 'condition'). The mind and body in ward settings
is manipulated and regimented, not only with bio-chemical treatments which
alter the very chemical constitution of the body, but also by a ward routine
which seeks continually to 'place' the body, geographically and temporally. As
Foucauldian analysis will tell us institutional regimes seek to discipline the body
in particular ways in particular spaces, and it is the resultant 'mortification of
self' (Goffman 1961: 24), the stripping away of different aspects of previous
social roles and (bodily?) characteristics (presumably coded as deviant in the
world outside the institution), which have prompted resistance in this case. The
Nottingham users, then, are attempting to change (in small, but significant ways)
these regulatory geographies of the body by critiquing the mundane practicalities of the ward (a tangible challenge) alongside, and as part of, the more difficult
and longer term project of introducing complementary medicines and 'talking
treatments' in order so that the mind and body are considered together. This
project signals a resistance to the notion and practices of the 'total institution'
(Goffman 1961) and to the inevitable disciplining of the mind/body in such
places.

Conclusion

In traditional inquiries into the geographies of mental health and illness, the individual is often overlooked as a source of information about health and place,
and relegated to the status of unproblematised geographical units. Recent writings within this subdiscipline have highlighted the need for more nuanced and
sensitive accounts of the ways in which health and place are meshed (Kearns
1993). This chapter has sought to contribute to this research agenda, and to
connect with other work emphasising the myriad of geographies which a 'post-
medical' geography might consider, including those of the mind/body (Dorn
and Laws 1994). My case study is written through to emphasise that patients of
psychiatric services do not have to be seen always as helpless subjects of medical
surveillance and control, and that meaningful processes of resistance to medical
categorisation do take place, even within medical–institutional geographies.
Spaces of resistance for psychiatric service users in Nottingham traverse the
mind/body, the institution and the community, and different social relations are
involved with and between each. These social relations are intimately tied up

with the contested but increasingly acknowledged status of mental patients as active agents, capable of advocacy for others and political action, and involved in differential resistance to medicalised identities. The particular place of Nottingham, its history as a centre of innovative mental health services (Parr and Philo 1996), and the recent organisation of the city's service users combine as interrelated factors which shape a complex 'dialectic' of health and place. Although this dialectic has not been fully excavated here (but rather over a series of works: Parr 1997a, 1997b; Parr and Philo 1995; Parr and Philo 1996), user critiques hinging on the mind-and-body emerge as one particularly important factor in the contemporary geography of mental health in Nottingham, and may have implications for other places and other geographical studies.

The body, or more specifically, 'mind and body space' (as outlined within the introduction to this volume) can perhaps be seen as a curiously neglected geography of mental health. This neglect may be related to the fact that geography has traditionally separated understandings of minds and bodies, and rarely considered them together. Individuals with mental health problems have traditionally been seen as collections of indicators such as diagnoses, dots on maps, or disembodied threats to suburban communities. These conceptualisations are perhaps representative of geography's reliance upon constructs of rationality and dualistic knowledge: 'Rose argues that Reason is not the whole story of masculism; in order to establish rationality, there must be a contrast with the irrational. Disciplinary knowledge can define itself through its own ability to know only if there are others who are incapable of knowing' (Longhurst 1995: 99). This infers (on two levels) that geography analyses the lives of individuals with mental health problems through constructs of rationality by (a) mapping their lives in ways which reveal limited dimensions to everyday life, and hence seeing the bodies as merely products of disordered minds, and by (b) not treating them as embodied, capable agents with political agendas which intersect with wider social processes. The marginalisation of people with mental health problems through ignoring multiple dimensions of their everyday experiences and subjectivities can be seen as an 'othering' process within the discipline. In order to counter this limited view of the lives of such people with mental health problems, emphasis on the body has been an important pathway to understanding both embodied political action and the lived experience of mental health problems. The mind/bodies of psychiatric patients are contested sites of control over which medical practices prescribe (and inscribe) alteration and correction. The collaborative geographies of resistance constituted by the user movement occasionally intersect with the site of the mind/body as a discursive (and actual) battle ground. It is important to recognise this, and to write it through as part of a geography of mental health. As Longhurst argues: 'constructionist feminists tend to be concerned with the processes by which bodies are written upon, marked, scarred, transformed or constructed by various institutional regimes'

(1995: 101). If we see minds/bodies as spaces which are inscribed in this way, we can further critique uses of medication as contributing to the social situation of the psychiatric patient. Here the mind/body as a location – mobile, but marked – replaces the asylum as a container of stigma and particular medicalised social relations.[12]

In an age of community care, where psychiatric patients are making lives in everyday spaces, mobility can be seen as resistance, with 'ill' or 'mad' minds/bodies claiming space within the city for expression and identification (Parr 1997b), and thus subverting notions of the body as a bounded medical site over which the owner has little control. The erratic yet continuing attention to psychiatric medication and possible complements/alternatives by the user movement in Nottingham is contributing to this disruption of traditional ideas about mental patients. What I have tried to show here is that psychiatric service users are not oblivious to trajectories of social and medical power, and are actively seeking to contest, resist and negotiate such relations. In resisting such power and categories, there is a return to ideas and spaces of 'nature' (in respect to projects such as Ecoworks) and to 'natural properties' (e.g., herbal therapies), albeit in conjunction with 'talking treatments', but what is refreshing about this is that these concerns seem to be arising from users themselves and not from any externally imposed medical and moral discourse.

Acknowledgements

I would like to thank NAG and NPCSG for allowing me to carry out my Ph.D. research. Brian Davey and Brenda Alexander were particularly helpful with materials and conversations relating to the concerns of this chapter. I would also like to thank Chris Philo for insightful comments and questions, and also Fiona Smith and Ruth Butler for casting a useful eye over the final product.

Notes

1 In a recent debate about geographies of disability, Golledge (1996) criticises Imrie (1996) and Gleeson (1996) for appearing 'politically correct' when questioning his use of medicalised terminology. As I have pointed out elsewhere (Parr 1997c), names, terminology and self-naming are important when considering issues of medical authority and individual and group resistance to medical categorisation.

2 Several users of mental health services in Nottingham who had experienced treatment at Saxondale Hospital (the now closed county asylum: see Parr and Philo 1996) grouped together to form a user council in 1986, and this initial group formation followed a World Mental Health Conference held in Brighton the previous summer. At the conference delegates from Holland had talked about user councils and advocacy, and these ideas were brought back to Nottingham by two mental health workers (Barker and Peck 1986).

3 Interestingly, Jones was a consultant psychiatrist at Mapperley Hospital in Nottingham, a key site for user organisation and influence.

4 The recent rise in consumption of herbal medicines for mental health can be interpreted as 'capitalist neurosis' in different ways. Firstly, as a commodification of depression and stress as capital diversifies from the mass manufacturing of pharmaceutical medicine to mass manufactured herbal remedies in an attempt to saturate the market (with respect to lay beliefs and actions about health and illness). Secondly, this could be interpreted as a response to the stressful conditions which contemporary time–work disciplines and spaces of advanced capitalism impose upon the mind/body. Thirdly, a more positive interpretation supposes that alternatives to conventional medicines offer a more potentially liberating and empowered pathway to (self-administered) treatment for different mind/body states.

5 'Talking treatments' refer to therapies which engage with an individual's psyche and which effectively deny the usefulness of the totally body-based explanations/treatments of medical psychiatry.

6 Such choices, it is noted in the psychiatric literature, are commonly represented as part of the role of being a 'bad patient' (Estroff 1987).

7 Brian Davey, the Nottingham Advocacy Group development officer, is a key figure within the organisation and has written various articles about the politics of mental health (Davey 1988, 1993, 1994).

8 These are a range of alternative/complementary therapies which are mediated through a variety of practical applications and conceptual bases. These vary from the physical manipulations of the body (through massage and needles) to the use of scented oils (on and surrounding the body) to the administering of different 'natural' remedies (orally) which claim to help alleviate a wide range of ailments.

9 Although not expanded upon here, these 'contexts, circumstances and settings' refer to how 'alternative' body therapies are or can be combined sensitively with more conventional therapeutic encounters, in particular rooms and buildings (ones potentially more empowering than the ward or psychiatrist's office). Ethnographic and interview work with this group of users has revealed their concern for changing the power relations of the therapeutic encounter through a changing of the environments of mental health care, hence the quotation on page 187 concerned with finding different ways to help people 'get back on their feet in new settings' (Interview with NAG worker 14 July 1993).

10 The community project of Ecoworks is a scheme which provides opportunities for voluntary activity, training and employment for people with mental health problems, and the underlying philosophy of the project explicitly links mental health with active individual and collective participation in social and physical environments. Primarily the emphasis is on a 'green environmentalism' that provides a psychologically healthy local alternative or complement to a perceived disempowering mental health system. Although too much to discuss in detail here, the connections that have been fostered in Nottingham between environmentalism and conceptions of citizenship are testament to the diverse foci of user-led initiatives seeking alternatives to traditional scientific explanations of the self. Gardens and gardening used to be a large part of in-patient life in the old asylums, especially ones that were also self-sufficient farming communities. To disrupt these histories of often enforced 'therapeutic' labour is another example of how users are reappropriating places and practices which were traditionally the domain of the mental health professional.

11 In addition to the other somatic treatments indicated earlier in this chapter, there are other controversial and gruesome 'therapeutic practices' such as ECT and lobotomisation which have been particularly important within twentieth-century psychiatry (see Jones 1983: Chapter Four).

12 It could be argued that the body is also a site of mad/ill inscription through practices of self-harm. Although these practices often result from individual trauma, the practice does serve to reaffirm control over one's own body.

References

Alexander, B. (1993) 'The place of complementary therapies in mental health', unpublished report, Nottingham Patient Councils Support Group.

Barker, I. and Peck, E. (1986) 'Movement in Holland', in The Nottingham Patient Council Support Group Information Pack.

Bell, D. and Valentine, G. (1995) *Mapping Desire: Geographies of Sexualities,* London: Routledge.

Bell, D., Binnie, J., Cream, J. and Valentine, D. (1994) 'All hyped up and no place to go', *Gender, Place and Culture* 1: 31–47.

Butler, R. and Bowlby, S. (1997) 'Bodies and spaces: an exploration of disabled people's experiences of public space', *Environment and Planning D: Society and Space* 15: 411–33.

Cream, J. (1994) 'Out of place', paper presented at the Association of American Geographers Conference, San Francisco, April.

Davey, B. (1988) 'Users councils in Nottingham', *Asylum: A Magazine for Democratic Psychiatry* 2: 8–10.

—— (1993) 'The place of complementary therapy in mental health', unpublished paper, Nottingham Advocacy Group.

—— (1994) 'Mental health and the environment', *Care in Place* 1, 2: 188–201.

Dean, K. G. (1979) 'The geographical study of psychiatric illness: the case of depressive illness in Plymouth', *Area* 11: 67–171.

Dorn, M. and Laws, G. (1994) 'Social theory, body politics and medical geography', *Professional Geographer* 46: 106–10.

Duncan, N. (1996) *Bodyspace: Destabilising Geographies of Gender and Sexuality*, London: Routledge.

Dyck, I. (1995) 'Hidden geographies: the changing lifeworlds of women with multiple sclerosis', *Social Science and Medicine* 40, 3: 307–20.

Estroff, S. E. (1987) 'No more young adult chronic patients', *Hospital and Community Psychiatry* 38, 1: 5.

Gesler, W. (1992) 'Therapeutic landscapes, medical issues in the light of the new cultural geography', *Social Science and Medicine* 34, 7: 735–46.

Giggs, J. A. (1973) 'The distribution of schizophrenics in Nottingham', *Transactions of the Institute of British Geographers* 59: 55–76.

Gleeson, B. J. (1996) 'A geography for disabled people?', *Transactions of the Institute of British Geographers* NS 21: 387–96.

Goffman, E. (1961) *Asylums: Essays on the Social Situation of Mental Patients and Other Inmates*, Harmondsworth: Penguin Books.

Golledge, R. E. (1996) 'A response to Imrie and Gleeson', *Transactions of the Institute of British Geographers* NS 21: 404–11.

Grosz, E. (1993) 'Bodies and knowledges: feminism and the crisis of reason', in L. Alcoff and E. Potter (eds) *Feminist Epistemologies*, New York: Routledge.

Illich, I. (1977) *Limits to Medicine, Medical Nemesis,* Harmondsworth: Penguin.

Imrie, R. (1996) 'Ableist geographies, disableist spaces: towards a reconstruction of Golledge's "Geography and the disabled"', *Transactions of the Institute of British Geographers* NS 21: 397–403.

Jones, W. L. (1983) *Ministering to Minds Diseased: A History of Psychaitric Treatment*, London: William Heinemann Medical Books Ltd.

Kearns, R. A. (1990) 'Coping and community life for people with long-term mental disabilities in Auckland', occasional paper no. 26, University of Auckland, New Zealand.

—— (1993) 'Place and health: towards a reformed medical geography', *The Professional Geographer* 45: 136–47.

Longhurst, R. (1994) 'The geography closest in – the body . . . the politics of pregnability', *Australian Geographical Studies* 32, 2: 214–23.

—— (1995) 'The body and geography', *Gender, Place and Culture* 2: 97–105.

Lowen, A. (1975) *Bioenergenetics*, Harmondsworth: Penguin.

McDowell, L. and Court, G. (1994) 'Performing work: bodily representations in merchant banks', *Environment and Planning D: Society and Space* 12: 727–50.

Merleau-Ponty, M. (1962) *Phenomenology of Perception*, trans. C. Smith, New York: Humanities Press.

Nash, C. (1996) 'Reclaiming vision: looking at landscape and the body', *Gender, Place and Culture* 3: 149–69.

Nottingham Patient Councils Support Group (1986) *Information Pack.*

Parr, H. (1997a) 'Sane and insane spaces: new geographies of deinstitutionalisation', unpublished Ph.D. thesis, University of Wales, Lampeter.

—— (1997b) 'Mental health, public space and the city: questions of individual and collective access', *Environment and Planning D: Society and Space* 15: 435–54.

—— (1997c) 'Naming names: brief thoughts on disability and geography', *Area* 29, 2: 173–6.

—— (1998) 'Mental health, ethnography and the body', *Area* 30, 1: 28–37.

Parr, H. and Philo, C. (1995) 'Mapping mad identities', in S. Pile and N. Thrift (eds) *Mapping the Subject: Geographies of Cultural Transformation*, London: Routledge.

—— (1996) *A Forbidding Fortress of Locks, Bars and Padded Cells? The Locational History of Nottingham's Mental Health Care*, Historical Geography Research Series No. 32.

Philo, C. (1992) 'The space reserved for insanity: studies in the historical geography of the mad business in England and Wales', unpublished Ph.D. thesis, University of Cambridge.

—— (1996) 'Enlightenment and the geographies of unreason', unpublished paper.

—— (1997) 'Across the water: reviewing geographical studies of asylums and other mental health facilities', *Health and Place* 3, 2: 73–89.

Pile, S. (1996) *The Body and the City: Psychoanalysis, Space and Subjectivity,* London: Routledge.

Porter, R. (1989) *The Social History of Madness*, London: Weidenfield and Nicolson.

Rodaway, P. (1994) *Sensuous Geographies: Body, Sense and Place,* London: Routledge.

Rogers, A. and Pilgrim, D. (1989) 'Mental health and citizenship', *Critical Social Policy* 26: 44–55.

Rose, G. (1993) *Feminism and Geography: The Limits of Geographical Knowledge,* Cambridge: Polity Press.

Scheff, T. J. (1966) *Being Mentally Ill: A Sociological Theory*, London: Weidenfield and Nicolson.

Scull, A. T. (1977) *Decarceration: Community Treatment and the Deviant — A Radical View*, New Jersey: Prentice Hall Inc.

Sibley, D. (1995) *Geographies of Exclusion: Society and Difference in the West*, London: Routledge.

Smith, C. J. (1980) 'Neighbourhood effects on mental health', in D. T. Herbert and R. J. Johnston (eds) *Geography and the Urban Environment Volume III*, Chichester: Wiley.

Szasz, T. (1961) *The Myth of Mental Illness*, New York: Harper and Row.

Tancredi, L. T. (1983) 'Psychiatry and social control', in L. Romanucci-Ross, D. Moerman and L. Tancredi (eds) *The Anthropology of Medicine: From Culture to Method*, Massachusetts: J. F. Bergin.

Turner, B. S. (1987) *Medical Power and Social Knowledge*, London: Sage.

Warner, R. (1985) *Recovery from Schizophrenia*, London: Routledge.

11

DOUBLE THE TROUBLE OR TWICE THE FUN? DISABLED BODIES IN THE GAY COMMUNITY

Ruth Butler

The London Gay Pride march is repeatedly led by the disability rights banner, but the disabled[1] contingent of the crowd are marginalised in the parties which follow it by 'the cult of "body beautiful" and celebration of glamour and glitz' (Corbett 1994: 345). The same ableist obsessions with the perfect body which have been noted by feminists as strongly affecting women's lives (Baker 1984; Wolf 1990), and by disability activists as heavily influencing disabled people's lives, run deep in the gay 'community'[2]. As a result many disabled lesbians, gay men and bisexual individuals (LGBs)[3] feel the need to attend Pride festivals, not to support gay rights, but to draw the LGB population's attention to their disabled members (Shakespeare *et al.* 1996). This chapter questions why, how and at what cost disabled people remain marginalised in the gay 'community'.

Recent work on marginalised groups, through flirtations with post-modernism, has drawn attention to differences within populations which have formerly been considered to have a single unifying identity. The misleading and unhelpful nature of dichotomies such as those of male/female, homosexual/heterosexual and disabled/able bodied have been problematised by an awareness of the numerous social, economic and political axes – class, gender, race, age, sexuality and (dis)ability, amongst others – which cut across such simplistic binary divisions and are themselves fluid concepts.

This work has liberated knowledge and produced fruitful theoretical controversies about populations' identities and how we may 'know' them (Gibson-Graham 1996). An appreciation of the complexities of people's embodied experiences, and the forces of power and resistance which create and recreate such experiences have been brought to light. For individuals whose 'different' perspectives on the world have previously gone unheard in an ableist, hetero-patriarchal society, such work promises an acknowledgement of their existence and the possibility of their experiences and concerns being raised in broader debates. However, such developments in theoretical debate have not always been welcomed. There are fears, particularly in political movements, that discussions of difference will bring about divisions and disunity, threatening their political objectives.

This chapter considers these issues in relation to the experiences of disabled LGBs. I first consider the nature of the roles expected of disabled LGBs, relative to their perceived, common representations and actual, multiple identities. I will draw on discourses about the body to consider how social expectations surrounding it have been constructed and continue to function. I will look specifically at (a) LGBs' need for a sense of belonging and commonality with those around them, (b) the related pressures upon them to 'pass' as what is socially considered to be 'normal' in ableist hetero-patriarchal society, and (c) their need for access to spaces and places where they can display their 'true' character. Second, I discuss the reactions disabled LGBs have to such expectations of them and their methods of coping with both their physiology and the numerous, complex, interacting demands society places upon them. I finally draw attention to the implications of an awareness of such experiences and processes for the wider disabled and LGB populations and their political fights.

The central importance of an individual's body, and its associated identities, to their lived experiences is now widely accepted. However, what parts biology and social structure, the self and others, play in controlling the body remains open to argument. The next section outlines just some of the manifestations of different social, economic, political and physiological controls on the body which interact to produce the complex embodied experiences of disabled LGBs.

Representations and expectations of bodies

An individual's behaviour and presentation of self in public space results from a combination of first their awareness of what they know or believe themselves to be, and second how they believe others to view them and hence what they believe to be expected of them (Goffman 1963). On the one hand, having pride in oneself means self-confidence to express oneself and value ones identity (Corbett 1994). Recognition and acceptance of both physiology and sexuality is central to anyone's self-esteem, self-respect and a conscious sense of being a man or woman (Morris 1989). These are issues which gay pride and disability politics have both underlined. On the other hand, however, whatever pride in themselves disabled LGBs may have behind closed doors, they are distinctly aware of society's reactions towards them. The surveillance of others upon any individual's behaviour has strong implications for their 'performance' in different spaces.

In a self-defensive manner we all tend to see ourselves as 'normal', and hence socially acceptable, and those we view as 'other', as deviant (Shakespeare 1994b). In this way we each distance ourselves from the undesirable margins of society and what they are believed to represent (Sibley 1995). Self-defensive, distancing processes produce a social structure where we are all measured against an unspecified yet apparently desirable 'norm'.

It should be noted that the 'norms' expected of people and which they may strive to attain are not constant. Identities are 'constructed and reconstructed over time and space' (Valentine 1993b: 239). The social, economic and political categories that an individual places themself in, or is placed in by others at any one time, will strongly affect the 'norm' with which others expect them to comply. Some categories are themselves considered more 'normal' and hence desirable than others. For example, essentialist argument suggests that the biological instinct to reproduce means heterosexuality and able bodies are 'normal'. In ableist hetero-patriarchal society images of heterosexual, able bodied couples are represented as having the desirable lifestyle to which everyone should try to conform. Constructionists have, more recently, argued that biological differences do not have any inherent meaning, but rather that characteristics and behaviours have been attributed to biological phenomena. However, what is common to both these arguments is the recognition of the impact of the 'norms', of biological or social creation, which individuals recognise and accommodate in their self-presentation and behaviour in order to survive the social jungle.

The need to belong

Fears of marginalisation, due to a lack of conformity, mean that the need to know that one is not alone in one's circumstances is important for any individual. This is a feeling recognised by Polio (1994) who recalls her efforts as an isolated, lesbian, disabled mother to contact others who might be in a similar position for support. Many LGBs upon recognising their sexuality suffer from loneliness and depression as isolated individuals outside the gay 'scene'[4] (Valentine 1993a). There is a high attempted suicide rate among young people who have identified themselves as homosexual and do not have the relevant knowledge or support to be able to deal with it (Valentine 1995). In this way a strong influence on the embodied experiences of a disabled LGB is their experience of contact with other like-minded individuals.

Having places where an individual can be open about their sexuality and enjoy the company of others with similar experiences is of great importance (Bell 1991). The gay 'scene' offers space to meet for support, companionship, to form relationships and to build a personal identity (Valentine 1993a). The 'scene' is particularly important to, and indeed dominated by, younger individuals who have not yet built up broader support and friendship networks. The lack of such spaces is a problem facing LGBs in isolated rural areas. The urban environment offers the benefits of both anonymity in a heterogeneous population with which it is easy to blend and a greater provision of services and facilities (Bell and Valentine 1995; D'Augelli and Hart 1987). However, a lack of access to the gay 'scene' for disabled people causes some LGBs difficulties even in the best serviced city.

Valentine (1993a) suggests that experiences of homophobia unites individuals across class, age and other social divisions, as mutual support fosters a sense of 'community'. All social backgrounds frequent the gay 'scene' it is argued, resulting in broad friendship groups, supportive of each others' common experiences. Whether the 'scene' is quite so tolerant in reality is questionable. It must in particular, be asked whether this 'united community' is inclusive of disability.

As a disabled lesbian, Field (1993: 18) notes that herself and others are sometimes 'not "invited" to be part of our own gay community'. She suggests that other LGBs often assume disabled LGBs do not exist and can certainly not conceive of having a relationship with such individuals. When disability is recognised it is often treated with the same patronising and humiliating tone that is common throughout Western society. A letter to *The Pink Paper* (Staples 1996: 10) highlights this point:

> I am collecting used postcards and used postage stamps in aid of the Guide Dogs for the Blind Association. This would go towards helping gay and lesbian blind people *who have very little in life* [emphasis added].

Crossing the divide and entering gay space involves a recognition of identity and can be a big step for any individual (Valentine 1993a). If the welcome is less than supportive due to ableism the experience can be all the more traumatic. Shakespeare *et al.* (1996) list many examples of ableism in the gay 'scene' collated from their interviews with disabled people. Examples involved: a man with a learning disability wearing a gay rights T-shirt being questioned as to whether he realised he was entering a gay bar; security staff at gay clubs repeatedly making excuses on the grounds of 'safety' for a disabled person to be barred from entering; people repeatedly being patronised or used as a tokenistic disabled friend at parties.

Invisible impairments may at first be considered less problematic, but false assumptions about 'ability' can lead to further misunderstandings and marginalisation. For example, if an impairment is revealed in an intimate social interaction the individual often suffers rejection and is aware of a rapid withdrawal by formerly interested parties (Shakespeare *et al.* 1996). Simple problems of an assumed understanding of communication through body language can cause a disabled person difficulties. The inability to make eye contact or stand up with ease can mean that finding a partner is no easy feat.

> Disabled people may . . . find it difficult to initiate contacts in pubs or at parties. To take the initiative and take a seat close to someone who is attractive may be very difficult for someone in a wheelchair, and for visually impaired people it may be impossible.
>
> (Oliver 1983: 72)

The inability to respond to another's body language when they 'take the initiative' may appear rude or give inaccurate suggestions that the individual is not interested. These are problems which disabled LGBs are aware of and which can make them feel as uncomfortable and isolated in gay space as all LGBs can be in heterosexual space.

Passing as 'normal'

The stigmatism and marginalised status of LGBs and the homophobic abuse which can occur can result in individuals attempting to hide their identity and 'pass' as 'normal', heterosexual, in spaces outside the security of the gay scene or behind the closed doors of the lesbian or gay household (Bell and Valentine 1995; Bell *et al.* 1994; Egerton 1990; Johnston and Valentine 1995; Valentine 1993a, 1993b).

It should be noted that problems of low income and discrimination in the housing market for some LGBs mean that the probability of home ownership and access to private space is reduced (Bell 1991). This is also a problem for disabled people as employment levels and access to insurance and credit services amongst them are low (Barnes 1991). For those individuals who fall into both groups the problems can be severe, compounding the importance of the gay 'scene' in which they can relax and be 'out' of both closets.

The practice of 'passing' in public space is a phenomenon that Morris (1989) recognises when noting the lack of information from lesbians in her study of disabled women's experiences of paralysis. She points to social pressure as the reason for their inability to speak freely. This need to conceal personal identity is not an experience of LGBs alone. It has also been recognised of disabled people, regardless of their sexual orientation. Corbett (1994: 344) writes:

> closeted gays pay too high a psychological price for passing, a contention that Abberley (1987) and Morris (1991) apply to disabled people.

An able body is seemingly preferable to impairment, which often reminds us of our physicality, animality, human frailty and ultimate death. Feminist writers have noted how women's bodies are often treated simply as objects (Young 1990), but it is not only women who suffer such treatment. People's fear of marginalisation encourage them to take on board ideas of the social 'norms' discussed above and build them into their own evaluation of their identity (Goffman 1963). There is 'a persuasive social pressure to be "more normal" than we are' (Corbett 1994: 346). We are all to some extent objectified by the gaze of others. Our efforts to shape our bodies into an image that both we and others are comfortable with in consumer culture have made the fashion and beauty industries highly profitable. Foucault suggests that power is diffused throughout all levels of the social order from national and international government to individual bodies (McDowell 1995). There are many forms of control, surveillance and discipline focused on the body constantly forming and reforming what are considered to be the appropriate 'norms' of behaviour and presentation. However, it is self-surveillance and self-correction which has one of the strongest influences in perpetuating a social structure which functions around such 'norms':

> There is no need for arms, physical violence, material constraints. Just a gaze, an inspecting gaze, a gaze which each individual under its weight will end by interiorising to the point that he is his own overseer, each individual thus exercising this surveillance over, and against himself.
>
> (Foucault 1977: 155, cited in McDowell 1995: 78)

Whilst it is still open to debate to what extent we choose to decorate our bodies to our own tastes and to what extent we feel forced to do so:

> Disabled men and women are encouraged by media representations of 'normal' bodies to obscure by dress and bodily decoration what are seen by others as bodily inadequacies.
>
> (Butler 1998: 86)

However, with many disabilities this is not an attainable goal. Many individuals have no alternative but to pay the price of obvious membership of the disabled population.

> Sexual confidence is so centrally about beauty, potency and independence that disabled women and men feel undermined.
>
> (Shakespeare 1996: 193)

'Ugly', 'deviant', disabled people are often treated as a single, homogenous group even though their circumstances in terms of both the individual's impairments and their social, economic and political circumstances can take many forms. The common assumption that all disabled people are asexual has largely gone unchallenged (Morris 1989, 1991; Greengross 1976; Oliver 1983; Corbett 1994; Shakespeare 1996; Shakespeare *et al.* 1996). It is an idea which the media establishment has been both influenced by and helped to reinforce (Shakespeare 1996). At times, images of sexual expression by disabled people have consciously been censored from public gaze. A case in hand was the attempted censorship by Westminster Council of the film *Crash*. Amongst the reasons for their desire to withdraw the film was a love scene involving a disabled woman (Norman 1996).

Fears of hereditary disease and images of incapable, weak, and generally undesirable impaired bodies make disability and asexuality seem synonymous in ableist, hetero-patriarchal society. The idea of disabled people as sexual beings is generally a source of either horror or amusement (Greengross 1976). Despite long running telephone sex lines and more recently the development of cybersex, sexual activity is still seen as a very physical, embodied experience which 'deviant' bodies cannot participate in satisfactorily. This is a theme picked up on by the cartoonist in Plate 11.1.[5] Homosexual activity has equally been questioned on the basis of physiology due to homosexual couples lacking the ability to have 'intercourse', in heterosexual terms, and hence what is considered a meaningful relationship. This arguably suggests more about heterosexual and/or able bodied people's fears of inadequacy than it does about disabled LGBs assumed inabilities.

Plate 11.1 Stupid question no. 154

However, despite their shared experiences of discrimination on the grounds of their assumed biological make up, the gay 'scene' offers little understanding to disabled people. Strong foci in the gay men's 'community' are the 'body beautiful', dancing and recreational drugs in which it is not always possible or desirable for disabled people to participate (Shakespeare 1996; Corbett 1994). Participants in Shakespeare's (1996: 201) research into disabled people's sexuality also reported a growing 'body fetishism' in the lesbian 'community'. It is acknowledged that many lesbians have chosen to challenge the social pressures for women to conform to male requirements of female bodily presentation, but it is equally noted that they can pay heavily for doing so. It is hard for any of us to meet the narrow ideals of physical beauty which exist in Western culture, yet the cost of marginalisation for those who deviate too far from the dream is high (Shakespeare 1996). Ableist views exist in both gay and heterosexual space and the pressures on disabled people to conform to the asexual role expected of them remain.

Access to the 'scene'

As with all areas of life, the movements of disabled people in the gay 'scene' are often limited by the physical structuring of the built environment (Shakespeare *et al.* 1996; Field 1993). Shakespeare (1996: 199) suggests that at venues 'where sex is on the agenda', the assumed asexuality of disabled people, as well as the repulsion of any idea of them as sexual beings, results in the failure to plan for disabled access to such spaces. This can be a self-perpetuating circumstance as inaccessible environments reduce the visibility of disabled people still further and any apparent need for the financial outlay to improve the situation.

Whilst it may be one of the most obvious, planning and building design are not the only ways in which the structure of the environment causes disabled LGBs problems. Due to the pressures of conformity and normalisation discussed above, 'in Britain, it is possible to be gay (only) in specific places and spaces' (Bristow 1989: 749). Availability of information in order to access such spaces is vital, especially with reference to isolated individuals in need of reassurance and support. However, a major problem in the lives of disabled LGBs is that a third party is often needed to gain access to information, people and places. As Shakespeare (1996) points out, 'professionals' may be prejudiced and insensitive over service provisions. So called 'professional' care workers, as well as parents and guardians, are as susceptible to homophobia and ableist attitudes towards disabled people's sexuality as any other member of society. To find personal assistants willing to support an individual in their desired lifestyle, especially if intimate assistance is needed before sex, is no easy task (Shakespeare 1996). The essential role played by 'matriarchal', 'network brokers' in lesbian friendship

Plate 11.2 Our rights, our lives, our choices!

networks (Valentine 1993a) equally relies on them being open to the idea of disabled people as sexual beings. The individual's rights to recognise and express their sexual identity, as depicted in Plate 11.2, are often denied.

Over-protective parents and close family members often cause the most serious problem of invaded privacy and restricted freedom of expression for disabled people.

> Parents of disabled youngsters are sometimes over protective and reluctant to allow their children to take the usual teenage risks. Furthermore, disabled teenagers may find it difficult to do things that perhaps they should not (when they go out they probably have to be transported by their parents). They therefore can't lie to their parents about where they have been or who they have been with.
>
> (Oliver 1983: 72)

This is a problem for heterosexual disabled youths, but homophobia amongst parents and carers adds another dimension to the difficulties of LGB disabled youths. LGBs with AIDS returning to the family home in rural areas may have to face the homophobia that caused them to move to the anonymity of the city in the first place (Bell and Valentine 1995). Carers, especially in cases of severe and or mental impairments, can control who the disabled person has contact with, at times restricting the access of long-term partners (Shakespeare *et al.* 1996). Whilst carers' intentions in this way may be good, they are often misguided by strong social beliefs which then become restrictive. Their love can be suffocating.

Accessing the safety and support of the gay 'scene' requires the possession of social and cultural, as well as material, capital. The importance of 'passing' to avoid homophobic abuse makes the ability to conceal ones sexuality a desirable skill. Flexibility in an individual's sexual image is advantageous at different times and in different places (Valentine 1993b). As a result, the ability to identify a homosexual individual in heterosexual space is not often an easy task (Valentine 1993a). 'Dyke spotting' is an acquired skill (Munt 1995). Learning to recognise the clues takes time for someone new to the 'community'. There are subtleties of dress which are employed to produce mutual recognition. Knowing looks, eye contact and other body language, and subtleties of architecture, layout and decor in gay pubs and clubs all help to keep the 'scene's' secrecy from the heterosexual world (Weightman 1980; Bell 1991). They need not be discussed in great depth here. Suffice to say that they exist, but that their significance relies heavily on an understanding of their visual symbolism and, therefore, the ability to see.

Word of mouth is another important way into the gay scene (Bell 1991; Valentine 1993a). Subtle references to homosexuality may be dropped into conversation to test others reactions in an attempt to make contact with other LGBs in heterosexual space (Valentine 1993a). However, a prior knowledge of the significance of given venues' names and other topics of conversation, as well as the ability to hear or lip read, are of great importance. It quickly becomes apparent how impaired individuals can be left isolated from such information sources and their rewards.

Switchboards and support centres are also an important means of access to the gay scene. Bell and Valentine (1995) point out the importance of anony-mous, accessible telephone services in the development of support networks between isolated individuals in rural communities. These services in urban areas are likewise of value. The gay press is also of significant value (Bell 1991). Advertisements for clubs, events or in the personal columns all have their obvious purpose if an individual has access to the relevant literature, television and radio outlets. However, these services often fail to cater for the needs of disabled people, particularly those with sensory impairments.

The invisibility of the disabled LGB population reinforces the lack of infor-mation in a suitable format for them. More publications in Braille and on tape for visually impaired individuals and more minicom systems on telephone helplines for deaf individuals are just two examples of what is needed. The lack of resources in the form of newspapers, books and other literature for isolated rural LGBs (Bell and Valentine 1995) mirrors the lack of information available to disabled people and limits the awareness of social and political advances among both populations. As Bell (1991) notes, gay bars and clubs open and close with high frequency and low profiles. Up-to-date information about their whereabouts is essential for access. This situation has been exacerbated by section 28 of the Local Government Act preventing the promotion of homosexuality by schools, youth group's and other government-funded establishments.

Fighting back

Whilst acknowledging the difficulties ableism, impairments and homophobia can cause it is not the intention of this chapter to paint a picture of doom and despair. It should be noted that most interviewees in Shakespeare's (1996) study of disabled people's sexuality accept their appearance and feel positive about their looks most of the time. Feminists' work on women's issues has drawn attention to their circumstances and aided their struggle to change those circumstances. Much disability research, however, even working within a feminist framework, has not empowered the disabled population in the same way, but passively illuminated a sorry state of affairs (Morris 1996). This is a pattern which I do not wish to add to. It would be inaccurate and politically dangerous to suggest that disabled people are either helpless or unresourceful. It must equally be remembered that a degree of initiative is needed by any individual attempting to access the gay 'scene' (Valentine 1993a). Fear of homophobic abuse, a lack of knowledge about or dislike for a venue, or restricted access in terms of transport can put limitations on any LGB's freedom of expression (Valentine 1993a). Disability must not be accepted automatically as either the sole or dominant cause of an individual's problems. It is not productive to consider being gay and disabled a 'double disadvantage' as one does not inevitably and continually compound the experiences, negative or positive, of the other.

However, the images we and others have of us do have to be addressed. To cope with social interactions, the expectations others have of us and in turn our views of ourselves, as discussed above, it is necessary to either put on an act accepting our expected role in society or publicly fight the expectations others have of us. Whatever our decision tensions will occur. It is difficult in homophobic, ableist space for a disabled LGB to display their true sexual and physical identity at any given time. The more commonly practised act of 'passing', equally comes at a price.

Whilst there may be little truth in others' assumptions about disabled people's sexuality, they can strongly affect their public identity, marginalised position in society and, when internalised, their self-image (Shakespeare 1996). 'Stigma, once established, is perpetuated and fuelled by internal oppression' (Corbett

1994: 345). To deny the existence of a person's sexuality restricts their ability to explore their identity and express themselves fully as already stated (Morris 1989). However, 'passing' as able bodied, can allow some individuals to express their sexual desires and still be taken as 'normal' at least some of the time, in gay space. For LGBs who cannot hide their impairments, others' misguided images of disabled people can have their advantages. For example, few people will think twice about an individual carrying a white stick taking hold of the arm of a person of the same gender. One interviewee in Morris' (1989) survey of women's experiences of paralysis notes how the assumption of asexuality placed upon her has removed the pressure on her to marry, take the role of a wife and mother and made it easier for her to live her life as a lesbian.

On a further beneficial note Shakespeare (1996) notes how sexual relationships involving a disabled person can put less pressure on both parties to 'perform'. He suggests that with the need to experiment, the emphasis can be moved away from Western society's obsessions with penetration and on to other parts of the body, the sense of touch and so on. As well as offering 'safe' sex this in turn results in a more equal power balance in the relationship. The nature of their impairment may mean disabled people have to take more responsibility for their sexuality and articulate what they want and need more effectively, thus leading to a better sexual relationship. It is possible to argue that disabled people may be more attuned to their bodies and feelings than able bodied individuals. The potential pleasures of experimentation are clearly illustrated in Plate 11.3.

Plate 11.3 If at first you don't succeed, try a new position!

For some there are times when complying to others expectations of behaviour and appearance is unacceptable and they choose to fight the images and expectations of them in a stronger manner, most obviously by attending pride festivals, proudly declaring their identities as 'Queer Crip's'. A growing literature on disability and homosexuality of both an academic and a more accessible nature equally has its political purpose (see, for example, Polio 1994; O'Toole 1996; Shakespeare *et al.* 1996; Tremain 1996). This type of more direct action has started to promote new images of disabled LGBs. It has began to draw the gay community's attention to the realities of disability. Role models are beginning to appear for other disabled LGBs. Such actions play their part in social education, integration and changing attitudes, offering hope to both disabled people and able bodied partners who, it should not be forgotten, also bear the stigma of disability, as they are often depicted as gold diggers, saints, insecure, inadequate individuals, or quite simply desperate (Morris 1989, 1991).

The problems facing disabled LGBs are being eased by general improvements in telecommunication, email, gay helplines and mainstream media services for the LGB populations as a whole. It is equally true that like-minded disabled people are learning to help themselves. Organisations such as GEMMA (a disabled lesbians organisation), REGARD (a disabled lesbian and gay group), VIGG (the Visually Impaired Gay Group) and LANGUID (Lesbians And Gays United In Disability) have been established to offer support and help between disabled LGBs. It should be noted that these support services are predominantly London-based at present, but their size, numbers and geographical dispersion are increasing. There is still, arguably, a lack of recognition of the differing experiences and needs of individuals with different impairments, social, economic and political circumstances, but disabled LGBs are, like other minorities before them, beginning to recognise their oppression and learning to fight back. The nature and implications of that fight back are the issues to which researchers and activists alike must now turn their attentions.

Conclusions: implications for the disabled and gay populations

There have been many similarities between the oppression faced by disabled people, regardless of sexuality and LGBs, regardless of corporeality. For example, all have been discriminated against on the basis of 'medical diagnosis' and both couples with disabled members and same-sex couples are met with social disapproval (Shakespeare 1996). In recent years the issues of HIV and AIDS have arguably increased the potential convergence of disability and gay politics (Corbett 1994; Hearn 1991). Yet as outlined in this chapter the 'gay scene' is not as understanding as may be expected (Corbett 1994). Whilst safe sex and the raising of funds for AIDS charities have relatively high profiles, access to gay venues for people with AIDS showing visible impairments is heavily restricted (Shakespeare 1996). The common experiences and needs of people with HIV/AIDS and people with other disabilities have not been recognised. HIV/AIDS is both a disability issue (Shakespeare 1994a; Campbell 1995) and a gay issue which could potentially unite the two overlapping 'communities' in their common struggles over oppression. There is little evidence to date to suggest that this is the case.

The significance of, and value associated with, a united front is connected to a belief in political strength in numbers and the old adage that 'together we stand, divided we fall'. Whilst essentialists and constructionists may disagree as to whether biological characteristics have any inherent meanings in an individual's embodied experiences, what remains common to both arguments is their support of a single unified identity of any particular oppressed population. Recognition of differences within and between the disabled and LGB populations have resulted in the proposal of policies of strategic essentialism (Gibson-Graham 1996). Whilst recognising that one would ideally want to challenge simplistic dichotomies Barrett (1991: 166) argues that 'political silencing can follow from rejecting these categories altogether', division and the loss of a group identity is apparently too costly.

I would suggest, however, that the dangers of failing to recognise populations' internal differences have been underestimated. Bell (1991: 323) stresses the importance of acknowledging the 'multiplicity of "gay geographies"'. He points to the different experiences of gay men and lesbians and notes that:

> Work must move away from emphasis on select 'gay Mecca', and researchers should be aware that findings are not fully transferable across space, time, gender, lifestyle. The 'gay community' must thus be seen in its full diversity, with studies focusing on single groups, or on certain spaces: there are different gay geographies of living, working and relaxing.
>
> (Bell 1991: 328)

Until difference is recognised the range of needs and services required to meet them will not be acknowledged or provided. The pain of marginalisation and exclusion will continue to be reality for LGB disabled people in the LGB and disabled 'communities'.

A Foucauldian perspective suggests that power is everywhere inscribed. From this standpoint Gibson-Graham (1996) argues that all work has a theoretical and political stance or entry point. One is not less political because it promotes difference. The politics of group identity is not the only viable political form. The modernist concept that knowledge and theory are separate and prior to change and

216

politics must surely be a fallacy as knowledge and its production is in itself a political process. There is no single knowledge which we all see and agree with, a single political aim for change.

Difference in this context is a political weapon which will be used either for or against social movements. Failure to recognise individual experiences as needs will itself result in division and disunity. Members of the populations that political voices seek to represent feel misrepresented, if represented at all, and more distant observers whom the movements seek to influence will be aware of its short falls from personal observation.

What is more, if implemented, the policies they strive for will be of little use to those at the grass roots level, whose needs have been ignored, but will delay the passing of further more practical legislation.

Geographers concerned with the issues of sexuality and disability have been slow to see the parallels and differences between the overlapping populations they study (Chouinard and Grant 1995 being a notable exception). Research into the complexity of people's lives is urgently required to develop a deeper understanding of society which can be put to practical use at all scales, ensuring that national and international civil rights legislation is of practical value to the everyday needs of isolated individuals. The effects of differing impairments, sexuality, race, gender, class, age and other variables on any group of people's lives must be considered. Only through doing so will a clearer understanding of the power relations at work in all areas of Western society be developed and ultimately broken down.

Acknowledgements

I would like to thank Hester Parr and especially Gill Valentine for covering earlier drafts of this chapter in helpful and thoughtful scribbles. Thanks also to Hester and Sarah Carty for their patience with my disregard for deadlines.

Notes

1 The following definitions of disability and impairment will be used throughout this chapter: impairment – lacking all or part of a limb, or having a defective limb, organism or mechanism of the body; disability – the disadvantage or restriction of activity caused by a contemporary social organisation which takes no or little account of people who have physical [or mental] impairments and thus excludes them from the mainstream of social activities (UPLAS 1976).

2 It is acknowledged that the term 'community' is one of the most elusive and vague terms in the social sciences. It has many different meanings to many different individuals. In this chapter I use it loosely to refer to the population of LGBs who have a broad sympathetic association, living not necessarily in the same area.

3 LGB is used throughout this chapter as an abbreviation for lesbian, gay men and bisexual individuals, due to the limited space available. However, it is stressed that these people are individuals and not a homogeneic unit as the abbreviation may regretfully infer.

4 The term 'scene' is used to refer to spaces and places (physical or other spaces of communication, such as cyber space) where LGBs can be out and express their sexuality freely, e.g., gay pubs and clubs, homosexual households, support groups, helplines, etc.

5 All the illustrations in this chapter are from 'Young Disabled People Do It Too', a series of cartoons which explore issues around, youth, sex and disability. They were produced in 1998 by the Young Disabled People's Project at DIAL House Chester, in conjunction with Cheshire Health Promotion and cartoonist Angela Martin. The images are available to buy in either postcard or poster format (cheques payable to 'YDP CHESTER') from: Young Disabled People's Project, DIAL House, Hamilton Place, Chester CH1 2BH, Tel: 01244 315025 (e-mail: ydpchester@compuserve.com). Set of five postcards £1.00 (inc p + p)*. Big 'A2' poster £2.00 (inc p + p)*. NB* overseas orders please add £2.00 per order.

References

Abberley, P. (1987) 'The concept of oppression and the development of a social theory of disability', *Disability, Handicap and Society* 2: 5–20.

Baker, N. C. (1984) *The Beauty Trap*, London: Piatkus.

Barnes, C. (1991) *Disabled People in Britain and Discrimination*, London: Hurst.

Barrett, M. (1991) *The Politics of Truth: From Marx to Foucault*, Cambridge: Polity Press.

Bell, D. (1991) 'Insignificant others: lesbian and gay geographies', *Area* 23: 323–9.

Bell, D., Binnie, J., Cream, J. and Valentine, G. (1994) 'All hyped up and no place to go', *Gender, Place and Culture* 1: 31–47.

Bell, D. and Valentine, G. (1995) 'Queer country: rural lesbian and gay lives', *Journal of Rural Studies* 11, 2: 113–22.

Bristow, J. (1989) 'Being gay: politics, identity, pleasure', *New Formations* 9: 61–81.

Butler, R. (1998) 'Rehabilitating the images of disabled youths', in T. Skelton and G. Valentine (eds) *Cool Places: Geographies of Youth Culture,* London: Routledge.

Campbell, J. (1995) 'Disabled people international', *UK Coalition of People Living with HIV and AIDS newsletter,* 7th edition.

Chouinard, V. and Grant, A. (1995) 'On being not even anywhere near "the project": ways of putting ourselves in the picture', *Antipode* 27, 2: 137–66.

Corbett, J. (1994) 'A proud label: exploring the relationship between disability politics and gay pride', *Disability and Society* 9, 3: 343–57.

D'Augelli, A. and Hart, M. (1987) 'Gay women, men and families in rural settings: toward the development of helping communities', *American Journal of Community Psychology* 15: 79–93.

Egerton, J. (1990) 'Out but not down: lesbians' experiences of housing', *Feminist Review* 36: 75–88.

Field, J. (1993) 'Coming out of two closets', *Canadian Woman Studies* 13, 4: 18–19.

Foucault, M. (1977) *Discipline and Punish*, London: Allen Lane.

Gibson-Graham, J. K. (1996) 'Reflections on postmodern feminist social research', in N. Duncan (ed.) *Body Space*, London, Routledge.

Goffman, E. (1963) *Stigma*, Englewood Cliffs, NJ: Prentice Hall.

Greengross, W. (1976) *Entitled to Love: The Sexual and Emotional Needs of the Handicapped*, London: Malaby Press.

Hearn, K. (1991) 'Disabled lesbians and gays are here to stay!', in T. Kaufmann and P. Lincoln (eds) *High Risk Lives,* Bridport: Prism Press.

Johnston, L. and Valentine, G. (1995) 'Wherever I lay my girlfriend, that's my home: the performance and surveillance of lesbian identities in domestic environments', in D. Bell and G. Valentine (eds) *Mapping Desire: Geographies of Sexualities*, London: Routledge.

McDowell, L. (1995) 'Body work: heterosexual gender performances in city work-places', in D. Bell and G. Valentine (eds) *Mapping Desire: Geographies of Sexualities*, London: Routledge.

Morris, J. (ed.) (1989) *Able Lives,* London: The Women's Press.

—— (1991) *Pride Against Prejudice*, London: The Women's Press.

—— (ed.) (1996) *Encounters with Strangers: Feminism and Disability*, London: The Women's Press.

Munt, S. (1995) 'The lesbian Flâneur', in D. Bell and G. Valentine (eds) *Mapping Desire: Geographies of Sexualities*, London: Routledge.

Norman, L. (1996) 'We have sex too', *The Guardian,* The Week 30 November: 6.

Oliver, M. (1983) *Social Work with Disabled People*, London: Macmillan.

O'Toole, C. J. (1996) 'Disabled lesbians: challenging monocultural constructs', *Sexuality and Disability* 14, 3: 221–36.

Polio, S. (1994) 'Being Sam's mum', in L. Keith (ed.) *Mustn't Grumble,* London: The Women's Press.

Shakespeare, T. (1994a) 'Disabled by prejudice', *The Pink Paper* 1 April: 13.

—— (1994b) 'Cultural representations of disabled people: dustbins for disavowal', *Disability and Society* 9, 3: 249–66.

—— (1996) 'Power and prejudice: issues of gender, sexuality and disability', in L. Barton (ed.) *Disability and Society: Emerging Issues and Insights,* London: Longman.

Shakespeare, T., Gillespie-Sells, K. and Davies, D. (1996) *The Sexual Politics of Disability,* London: Cassell.

Sibley, D. (1995) *Geographies of Exclusion,* London: Routledge.

Staples, D. (1996) 'Helping hand for the blind', *The Pink Paper* 17 May: 10.

Tremain, S. (1996) *Pushing the Limits: Disabled Dykes Produce Culture,* Toronto: Women's Press.

UPIAS (1976) *Fundamental Principles of Disability,* London: Union of the Physically Impaired Against Segregation.

Valentine, G. (1993a) 'Desperately seeking Susan: a geography of lesbian friendships', *Area* 25, 2: 109–16.

—— (1993b) 'Negotiating and managing multiple sexual identities: lesbian time-space strategies', *Transactions of the Institute of British Geographers* 18: 237–48.

—— (1995) 'Out and about: geographies of lesbian landscapes', *International Journal of Urban and Regional Research* 19, 1: 96–111.

Weightman, B. A. (1980) 'Gay bars as private places', *Landscape* 24: 9–16.

Wolf, N. (1990) *The Beauty Myth*, Toronto: Random House.

Young, I. M. (1990) 'Throwing like a girl: a phenomenology of feminine body comportment, motility and spatiality', in I. M. Young (ed.) *Throwing Like a Girl and Other Essays in Feminist Philosophy and Social Theory*, Bloomington: University of Indiana Press.

12

WITHOUT THESE WALLS: A GEOGRAPHY OF MENTAL ILL-HEALTH IN A RURAL ENVIRONMENT

Christine Milligan

Introduction

More than two decades have passed since Wolpert (1976) first drew our attention to the exclusion of those individuals who experience mental and physical disablement from mainstream society. More recently, human geographers have been called upon to examine the assumptions about inclusion and exclusion implicit within our social and spatial environment, and seek to identify forms of exclusion as articulated by marginal groups. Specifically, this chapter focuses on those aspects of the social and spatial environment that contribute to the inclusionary and exclusionary experiences of individuals with mental ill-health (MIH). The purpose of the chapter is threefold. First, it moves on from the early concerns of geographers with issues of deinstitutionalisation[1] to explore factors contributing to the geographies of a group of (normally) community-based individuals with MIH. Second, it discusses the ways in which inclusion and exclusionary practices are experienced by such individuals within the social and spatial environment which they are located. In doing so, it focuses on the issue of rurality and mental health, and third, it considers the response to MIH within the context of the Scottish rural environment. The aim is to highlight some of the key determinants contributing to the locational patterns of those with mental ill-health. Through the experiences of such individuals, we are also able to reflect on the ways in which the actions of these marginalised individuals can themselves contribute to the emergent geographies of MIH.

The development of community-based mental health care in Scotland

It is not the purpose of this chapter to discuss historical perspectives of MIH, as such issues have been well documented elsewhere (for a selection, see Scull 1977, 1979; Foucault 1979; Philo 1989, 1997; Murphy 1991; Prior 1993). Nevertheless, as Prior (1993) notes, representations of MIH are manifest in new texts, spaces, treatments, new occupational roles and social policies. Such representations are both contextualised by and contingent upon the past. Thus, as recent responses to MIH have not been developed in a vacuum, a brief discussion of factors contributing to the contemporary Scottish context are provided here.

Throughout the nineteenth and early twentieth century, care for those with MIH in Scotland, in common with provision in many other Western societies, was provided within walled and often isolated institutions. It is in the segregation of individuals with MIH within these 'sites of insanity', that it is possible to see the laying of a foundation for not only the spatial, but also their *social* exclusion from mainstream society. Though new clinical treatments developed over time, these institutions continued to represent the main sites of caring until the mid-1950s. The development of psychotropic drugs in the mid-50s however, combined with increasing criticism of the curative value of the asylums throughout the 1960s (e.g., Goffman 1961), led to an emergent belief in the need to move away from spatial exclusion towards domiciliary and community-based service provision. Within the British context, this newly emerging philosophy of community care was endorsed in the 1962 Hospital Plan, which launched the official closure programme of contemporary psychiatric hospitals. Official policy since 1962 has consequently incorporated the notion that care should be organised outside the traditional single-site, institutional environment, towards the development of small, geographically dispersed sites within the community. This momentum culminated in the 1989 White Paper and the subsequent 1990 NHS and Community Care Act. Emphasis has been on the reduction of residential care, with the aim of enabling those who need supportive care to 'live as independently as possible in their own homes or in a homely environment in the local community' (HMSO 1989: para.1.8). The social and spatial exclusion of those suffering MIH from mainstream society has consequently been replaced by inclusionary ideals manifest in new structures such as domiciliary-based support, group homes, community-based hostels and day centres, and a newly emergent geographical pattern of service provision.

In Scotland, however, the distinct legal system requires a separate policy amendment to legislation existing in England and Wales (Murphy 1992). The pace and form of mental healthcare development has of consequence varied by degree to that evident elsewhere in Britain (Pullen 1993; McMenamin 1996).[2] The 1985 report *Mental Health Care in Focus* described mental health services in Scotland as 'a deprived area of care' (Scottish Home and Health Department

1985: para.1.4), noting 'a serious shortfall in Scotland of community alterna-tives to in-patient mental health care' (ibid.: para.1.5). Nevertheless, whilst Scotland has not had a wide-reaching policy of moving psychiatric patients out of hospitals, community care for those suffering MIH has advanced spontaneously. Pullen (1993) maintains that the move of resources from institution to commu-nity has been unplanned, coming from local initiatives rather than Scottish Office policy. The Scottish Borders,[3] for example, implemented the first completely open-door psychiatric hospital in Britain as early as 1949 (Warner 1989), further adopting a model of home visiting and treatment as a means of servicing its widespread population in the 1960s. The Crichton Royal Psychiatric Hospital in Dumfries and Galloway has also operated a community re-integration programme for over twenty years. Consequently, in addition to the socio-temporal varia-tions identified by Prior (1993), regional and localised responses to MIH have contributed to variations in the social and spatial expression of mental ill-health.

Geographical perspectives

Deinstitutionalised care, in compelling society to change in a number of ways regarding issues that previously had little or nothing to do with those with MIH (i.e., local zoning decisions, housing, nursing home care, welfare benefits, etc.) purports to be inclusionary (Lewis *et al.* 1989). The lived reality of such an inclusionary system, however, has been questioned by geographers for over two decades (see, for example, Wolpert and Wolpert 1976; Smith 1981; Dear and Taylor 1982; Hall and Taylor 1983; Dear and Wolch 1987; Smith and Giggs 1988). Focusing largely on issues of deinstitutionalisation, such studies highlight the exclusionary practices of both society in general and local communities in particular. Combined with inadequate community-based alternatives, and the desire of some former long-stay patients to seek supportive networks largely found in inner-urban areas, it was demonstrated that a new geography of MIH was becoming manifest in a particular spatial form, the 'service-dependent ghetto' (Dear and Wolch 1987: 8).

The majority of those now being discharged from psychiatric hospitals, however, are no longer 'long-term', but 'revolving door' patients; a term used to describe those who experience recurrent periods of MIH. As hospitalisation for such individuals tends to be of repeated, but limited duration, they can often retain some familial or community ties. The 'revolving door' individual now forms the core of psychiatric in-patient re-admissions in Britain (MIND 1989). As Bachrach (1989) notes, the deinstitutionalisation process has effectively created a generation of 'new chronic'[4] individuals with MIH, who experience unique problems in accessing care. Such individuals are situationally different from their earlier counterparts as a result of service system changes. Consequently the factors contributing to the locational geographies of those with MIH as identified in earlier studies, may now be subject to shifting influences. As Lewis *et al.* (1989: 174) aptly comment, 'We have deinstitutionalized the patients . . . Now we have to deinstitutionalize our thinking and research, so that we can catch up with the patients in the new inclusionary system.'

It is important therefore, that deinstitutionalisation is not viewed as simply a change in the locus of care, but that we are cognisant of the ways in which the individual and the environment (e.g., the social setting in which the individual resides and interacts) can create inequities. Hahn (1985) comments that in order to achieve any full understanding of such issues we must assess not only environmental factors, but also the opinions and perceptions of disabled people themselves – that is, *their* assessment of the problems they encounter in their everyday lives. Whilst his writing refers primarily to those with physical disabilities, none the less it is of equal relevance to those with MIH.

To deinstitutionalise our research requires an examination of factors that influence the spatial geographies of Bachrach's (1989) 'new chronic' individuals with MIH. The locational preferences of these individuals, whilst in part shaped by new sets of social and environmental influences to those previously impacting on former long-stay patients, may nevertheless be influenced by the policies and practice of formal and informal service providers. By contributing to the support networks of those suffering MIH, such agencies are influential in the development of 'inclusionary', community-based systems of care. Thus, following a brief discussion on the issue of rurality and mental ill-health, the remainder of this chapter highlights through a small-scale study undertaken within the Scottish rural environment, the ways in which such actors and agencies can contribute to these newly emerging geographies of mental ill-health. This is demonstrated through an analysis of interviews with knowledgeable actors from the community, statutory and voluntary sectors, as well as group discussions and interviews with rural-based service-users.

The issue of rurality

Blank *et al.* (1995) have noted the historical neglect by both researchers and policy-makers surrounding the area of rural mental health. DeLeon *et al.* (1989) in particular notes the failure of policy-makers to differentiate between the needs of those with MIH in rural locales and that of their urban counterparts. As a consequence, there is relatively little empirical work to guide practice. Whilst the literature is scarce, some researchers, largely within the North American context, have begun to address these issues (e.g., Human and Wasem 1991; Hutner and Windle 1991; Sullivan *et al.* 1996; Kane and Ennis 1996). There is, however, little equivalent research in the British context. Those few rural studies that do exist have been largely undertaken by mental health professionals (e.g., McCreadie 1982; Robertson and McCreadie 1991; McCreadie *et al.* 1997) whose analysis is largely quantitative, and whose concern is primarily with clinical and biological models of MIH, rather than the influence of social or environmental factors. Such factors, however, must inevitably be of import in the development of any inclusionary system of mental healthcare. Whilst geographers have gone some way towards addressing social models of MIH, their focus has largely been on the urban environment (e.g., Giggs 1973; Dean 1979; Eyles 1986; Giggs and Cooper 1987; Moon 1988; Parr 1997). Yet geographers are particularly well qualified to address the social and environmental interactions of marginalised groups in rural locales. These are issues which this chapter goes some way towards addressing.

Whilst it is acknowledged that definitions of the rural are both contested and diffuse, most are defined on the basis of geographically widespread and sparsely populated areas and a preponderance of lower order settlements. Though most of Scotland's population live in or proximate to cities or towns, nevertheless, an important minority live in such geographically widespread areas. Recent work by geographers such as Philo (1992), Sibley (1994), and Cloke and Little (1997) has increasingly drawn attention to the need to consider a diversity of 'others' within the rural environment. Figure 12.1 highlights the area within which the research was conducted. Dumfries and Galloway is a largely rural region in south-west Scotland. As demonstrated by the population data, Dumfries, Annan and Stranraer are the most populous settlements within the region; Dumfries being the major town.

The case study

In moving away from the earlier focus on the deinstitutionalisation of former long-stay patients, the focus here is on factors impacting on the preferences of 'revolving door' individuals. The study highlights some of the key determinants contributing to their locational patterns. In doing so, it considers first the influence of statutory and voluntary bodies, before reflecting on the ways in which the individual contributes to the formation of the geographical landscape of MIH. Whilst the small scale of the study means no generalisations can be assumed, nevertheless the impact of such locational influences should be key to any consideration regarding the development of the geographical landscape of future service facilities.

The statutory sector

Historically the provision of mental healthcare services in Scotland has been the responsibility of the National Health Service. Within Dumfries and Galloway, this

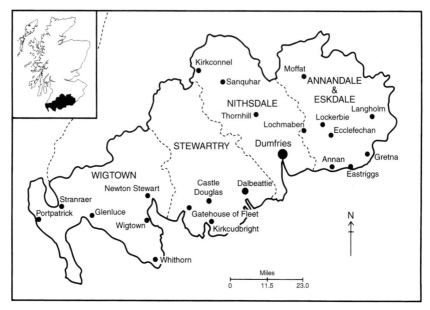

Population of Settlements within Dumfries and Galloway. (1991 Census Report)

Dumfries	37159	Sanquhar	2095
Stranraer	11348	Lochmaben	2024
Annan	8930	Eastriggs	1943
Lockerbie	3982	Thornhill	1633
Castle Douglas	3855	Wigtown	1117
Newton Stewart	3673	Whithorn	952
Kirkcudbright	3588	Gatehouse	909
Gretna	3149	Ecclefechan	847
Langholm	2538	Eaglesfield	642
Moffat	2342	Glenluce	579
Kirkonnel	2329	Portpatrick	536

Figure 12.1 Population of settlements within Dumfries and Galloway (1991 Census Report)

has centred around the Crichton Royal. Sited in Dumfries this is the sole adult in-patient facility for psychiatric care within the region. Bed numbers in line with current policy, however, have declined from around 1,200 in the 1960s to under 350 in 1994 and are projected to reduce to less than 70 by the end of the decade (Ferguson and Sanderson 1994). Clinics for the more general distribution of medication, however, are well dispersed geographically, though only one community day hospital catering for adult MIH exists within the region. This too is located in the urban core. As the current Dumfries and Galloway *Joint Community Care Plan (1995–1998)* (Dumfries and Galloway Regional Council and Health Board 1996) proposes no additional day-facilities for this group, Dumfries is likely to remain the focus of day hospital provision for the foreseeable future.

At present there exists no private provision for those suffering MIH, other than nursing homes providing for (largely) elderly dementias. Service facilities related to the medical welfare concerns of such individuals have consequently remained the remit of the statutory sector. With the exception of therapeutic support available at the community day hospital (which requires referral), there are no public sector *social* supports for those suffering MIH within the region. In terms of employment, the statutory sector operate one unit offering work placements for up to fifty people, though again this is located in the urban core. This contrasts sharply with ARC projects[5] which are well dispersed throughout the region.

The statutory sector has opted for an open and high-profile approach in its community-based siting strategies for mental health facilities. Planning proposals for the day hospital, for example, were accompanied by open meetings and letters to local residents in order to inform and encourage discussion of the proposal. However, as Grant (1992) notes, in Britain objectors have limited influence or legal recourse with which to challenge the locational decisions of statutory bodies. The centralised planning system means that local authority planning powers are closely defined by national legislation. Scottish Office guidelines (1994) state that whilst local communities should be *consulted*, this should not be confused with a right of veto.

The ethos and practice of community care, in moving away from the physical exclusion of those suffering MIH purports to be inclusionary. However, as earlier studies in the North American context have shown, local government electoral sensitivity to powerful neighbourhood opposition has enabled local communities to succeed in their efforts to physically exclude non-desirable service-dependent groups. Such action contributed to the concentration of those suffering MIH within the urban core, in effect creating an 'asylum without walls'. In Dumfries, however, despite strong and well organised opposition to the day hospital – including media coverage and the support of the then incumbent MP – the community were unable to prevent the siting of the facility.[6] Whilst local authorities may recognise the benefit of positive local support in integrating service-dependent populations into the community, this in no way confers any right of determination. Yet despite the limitations of the power of local communities to exclude *physically* those suffering MIH in the Scottish context, to date, the urban core has remained the favoured location for statutory facilities. A senior figure within the Health Authority[7] attributed this in part to the ability of locally powerful clinicians to influence the decision-making process, and who are reluctant to decentralise their power-base. Furthermore, as Kane and Ennis (1996) note, there are often difficulties inherent in recruiting mental health professionals to rural locations, where they may be geographically and socially isolated from their colleagues.

The voluntary sector

The new legislative framework for community-based care in Britain is aimed at shifting the locus of service provision from the statutory to the informal sector. As a result, the voluntary sector has an elevated role in the provision of services to community care groups.[8] The social and supported housing needs of those suffering MIH in Dumfries and Galloway, for example, are almost entirely the remit of the voluntary sector. Statutory bodies adopt an essentially enabling role, in the form of resource transfers, either through specific ring-fenced funding,[9] grant aid (or increasingly contracting), or the transfer of local authority tenancies to voluntary organisations for the provision of supported housing.

Contrary to the findings of earlier studies (e.g., Dear and Wolch 1987) those with MIH in this study were found to reside in a variety of neighbourhoods, in both public and private housing. This in part reflects the status of the 'new chronic' individual, who no longer experiences long-term institutionalisation. Despite short periods of hospitalisation, he or she may retain their community links, so returning either to the familial home or to proximate unsupported accommodation. The ability of local communities to exclude physically in these circumstances are consequently reduced. Supported housing is also well dispersed throughout the region, and found in both private and local authority housing providing for a variety of needs. Though a few individuals on leaving the rehabilitation unit have been discharged to either a hostel for young homeless (in Dumfries) or bed and breakfast accommodation, this is a declining form of provision. The remainder return either to the familial home or do not require supported accommodation, and so enter mainstream housing.

Supported accommodation has been limited to a maximum of two to three adults sharing, with the exception of three larger units in Dumfries housing up to five people. This strategy means that the purchase of housing becomes a personal transaction between the seller and the voluntary body. Registration of the property does not occur until *after* the purchase, thus the community remains largely unaware of the property's intended use until the sale is complete. This pre-empts the marshalling of any effective opposition. Further, by adopting a maximum occupancy level of five people, organisations side-step the need to apply for permission for change of usage under central planning regulations.[10] This has enabled them to locate housing opportunities for their client group in both public and private sector housing throughout the region, without subjection to those restrictions applicable to larger buildings requiring conversion. Such strategies have helped to circumvent the stigmatisation and attempts at physical exclusion commonly associated with the siting of larger units.

Voluntary agencies are also the main providers of social support and employment opportunities in the region. Outwith Dumfries, charity shops act as the core around which supported housing and a [very] few work opportunities are clustered. With the exception of the two most populous areas, however, voluntary sector drop-in centres – the basis for social support – operate on a part-time basis only. The statutory day hospital, though providing therapeutic as well as medical facilities, requires referral, and was in general seen as 'too formal' and 'too clinical' by facility users in this study. As a consequence, drop-in centres, in their opinion, were viewed as the only socially supportive environment *openly* available to individuals with MIH. The only readily available phoneline support to those undergoing a period of crisis was the Samaritans,[11] the voluntary sector whilst aware of this gap in provision, had insufficient funds with which to implement such a scheme beyond those currently in supported housing.

The voluntary sector can be seen to play a vital role in the delivery of community-based services to those with MIH. Their ability to act, however, is often limited by the inherent difficulties in attempting to implement effective models of social support in geographically widespread locales,[12] with limited resources, and their frequent reliance on statutory sector funding. Whilst community-based facilities may have improved, progress across Scotland is variable. Both at the inter- and intra-regional levels, variations exist in the availability of voluntary sector supports (Milligan 1998). This in part stems from variations in the historical development of the voluntary sector across Scotland. However, it is also the case that geographical variations in voluntary provisioning can arise as a result of differentials in local authority priorities and patterns of spending, contributing to a limitation of voluntary sector services, and a growth of geographical inequity in voluntary sector provision. As Bonthron (1995) notes, in some localities there is considerable anxiety amongst mental healthcare consumers with regard to service availability and the means of accessing services. For those located in sparsely populated areas, these issues can be exacerbated, contributing to what is effectively the exclusion of rural dwellers from community – but largely urban-based – facilities.

Whilst the policies and siting strategies of the statutory and voluntary sectors can be seen to be contributing to the emergence of a new local geography of MIH, the needs and experiences of individuals may counteract such strategies. Without the development of adequate social and employment opportunities, for example, the geographical dispersion of housing and medical clinics may be insufficient to alter the locational preferences of facility-users themselves. These issues are explored in this last section.

The role of the individual

Local geographies of MIH are also shaped by the interaction of individuals with their social and spatial environment. Here, these issues were explored through focus group discussions and interviews with 'revolving door' individuals with MIH located in various settlements throughout the region. These individuals were resident in both supported and unsupported accommodation, and all attended locally based clinics for regular medical treatments. Three focus groups were held in total, and seventeen adults with MIH, of mixed ages and sex, were interviewed. The experiences of 'Ian', 'Jim' and 'Paul',[13] summarised here, focus on the ways in which the locational geographies of MIH are influenced by the needs and preferences of service users themselves. The perception of stigma and heightened visibility in rural locales, social isolation, the conflict between housing opportunities and the concept of home, along with the importance of employ-ment opportunities, can all be seen to combine to create geographies of MIH that may be at variance with the policy and practice of service providers.

Ian's experience highlights the way that those with MIH in less populous areas can feel highly visible and socially isolated. Burvill *et al.* (1984), in a study of rural Western Australia, note the difficulties of maintaining confidentiality in small towns, and the stigma and social conspicuousness of those with MIH in rural compared to urban environments. Consequently, rather than experiencing greater integration into the community, individuals in rural areas can feel subject to a greater degree of stigmatisation. Exclusion is not expressed through the *physical* segregation experienced by earlier individuals with MIH, but by a felt social exclusion. Here, this is linked to 'Ian's' perception of stigmatisation,

---

Ian

Temporarily resident in halfway housing at the Crichton Royal, Ian had previously been allocated an unsupported flat in his home settlement (pop. 3,500) where he might benefit from familial support. He maintains, however, that his family have been unable to cope with his problems and have disowned him. He is unemployed, and no therapeutic employment opportunities are available in his locale. The sole social facility for those with MIH is a voluntary run drop-in centre, open two afternoons per week and attended by around ten people (in total). As a result, prior to his hospitalisation, he travelled forty-four miles a day to use facilities in Dumfries. He believes that small communities have a greater awareness of individuals with MIH. He feels labelled, stigmatised and socially isolated. He was willing to accept placement in homeless hostel accommodation (in the urban core) when discharged, despite his awareness that others similarly placed find this accommodation depressing and lonely. He believes Dumfries will give him an anonymity not found in smaller communities. Ian further believes that this is where he will find understanding friends and the social support networks he needs. He expressed his unwillingness to move into supported housing which he feels would further stigmatise him, preferring to live in 'ordinary' housing within the community, where he feels he would be able to lead a 'normal' life.

---

visibility and social isolation. His preference is therefore expressed for the anonymity of more populous areas, which provide the support facilities and social networks not readily available in more rural environments.

Ian's experience highlights an additional problem that may be faced by those with MIH in rural locales. Whilst the 1989 White Paper acknowledges the role of statutory and independent bodies in service provision to community care groups, it notes that in reality, 'most care is provided by family, friends and neighbours' (para. 2.3). Berry and Davis (1978) note, however, that family involvement is a varied experience, and whilst any setting where the family is hostile, apathetic or absent is difficult, for rural dwellers with MIH the problems can be exacerbated. With fewer acceptable alternatives in terms of lifestyle or employment opportunities to allow independence, the dependence on family is increased. For Ian, lack of familial support, combined with poor employment opportunities and social options, proved key 'push factors', culminating in his desire to leave the rural environment in which he resided to seek the support networks of the urban core. This occurred despite his awareness of the negative elements of urban life that he was likely to encounter.

The part-time provision of social facilities and poor employment opportunities in the more rural areas appear insufficient to prevent movement towards a more populous environment. So whilst housing providers may consciously adopt policies of dispersal aimed at avoiding agglomeration in the urban core, the desires of individuals with MIH can counteract these processes. In Jim's experience, though the desire for suitable housing resulted in relocation to dispersed housing alternatives, social networks are exerting strong 'pull-factors' overriding residential location.

For 'revolving door' clients who retained their social ties, the desire to locate proximate to the supportive networks of friends/family was a common (though not unanimous) factor. Most, however, sought not to return (where possible) to the parental home, but in housing sufficiently proximate to benefit from familial support. This appears to be due first, to a need to maintain a separate 'space' of their own, and second, because carers (generally parents) can find it difficult to cope with the stress and symptoms of long-term MIH. This awareness of possible stress to the relationship created by residence within the familial home can consequently manifest itself in the locational preferences of the individual.

Jim's experience highlights a further issue raised by Kearns and Smith (1994) in which they note that being *housed* is not the same as being *at home*; of equal importance is a sense of belonging. Community acceptance and responsibility cannot be taken for granted, the 'freedom to withhold and isolate' competes strongly with the freedom to include (Wolpert 1976: 10). Whilst local communities in the Scottish environment, for example, may have been less successful than their North American counterparts in spatially excluding those viewed as

Jim

Jim identified himself as a schizophrenic. He is in his late forties, and lives in shared, supported accommodation in a small town seventeen miles from Dumfries. He had previously been resident in a five-bedroomed hostel unit in Dumfries, but had been unable to cope with multiple residency. The location of residence was not his choice, but was the only small-scale supported accommodation available. He is at present sharing the house with another male with whom he was previously unaquainted. His co-resident's habits 'annoy him', he feels a lack of privacy and the need for his own 'space'. He has family in Dumfries, and feels if located there, he would gain from greater familial support. He has little to do with his neighbours, who whilst not unpleasant to him, quickly realised the significance of supported housing – emphasising the stigma still attached to MIH. Circles of friendship (he claims) are largely formed amongst those with similar problems.

'non-desirable' from their locales, communities still retain an ability to socially exclude. Both Jim and Ian highlight the issue of stigma and their perception of social exclusion from the community within which they reside. Thus as Jim commented, whilst individuals with MIH may be *in* the community, this in no way infers that they are part *of* the community.

This final example highlights the willingness of rural dwellers to travel considerable distances to satisfy the social or employment needs not met in less populous environs. This was a recurrent theme amongst interviewees. In this study, Dumfries was generally viewed as the hub of the social network for those with MIH. So while some interviewees did occasionally visit drop-in centres in other parts of the region, most travel was from rural locales towards the core.

Transportation presented several problems arising from: cost, rurality and the fact that some medications rendered users unable to drive, therefore reliant on family, friends, or public transport. As Murray and Keller (1991) note, however, the lack of public transportation, often taken for granted in urban environments, can interfere with the social interchange and accessibility of human services. As McCreadie commented, 'daycare is offered to all Nithsdale[14] schizophrenics . . . but distances and problems with transport in a rural area dictate that only those living in or around Dumfries can be reasonably expected to attend' (1982: 585). These issues remain unaddressed by service providers within the region.

Work opportunities were also a key factor affecting locational patterns. Voluntary agencies, the primary source of therapeutic work opportunities, acknowledged that

Paul

Paul is in his early thirties. He describes himself as having severe schizo-phrenia, and requires supported accommodation. Lack of familial support and community-based facilities in earlier years meant Paul spent twelve years in bed and breakfast accommodation in Dumfries. He was excluded from his accommodation between 10 a.m. and 6 p.m. He had no cooking facilities, and was required to eat separately from 'regular' clientele. The desire for his own home and privacy led him initially to accept supported accommodation in a small town seventy miles away, this move was unsuc-cessful. He then accepted similar accommodation in a village twenty-seven miles from Dumfries. He knew no one in either of these locations. In his first week in his second location he had money given to him for furnishings stolen by a new 'friend'. His social life, the friends from whom he draws support, and his therapeutic employment are all centred around Dumfries. His medication renders him unable to drive, yet despite the distance involved, at the time of interview, he was travelling a minimum of three times a week to Dumfries in order to continue working and maintain the social network he has built up over the years.

whilst they aimed to provide part-time employment to those with MIH, oppor-tunities were limited, and few associated with these groups had found full-time employment. Shadish *et al.* (1989) link the high rate of unemployment found amongst those with MIH, to the political and economic rationale inherent in the formation of social policy, which they argue is exclusionary to those marginalised groups who are not well represented in the market-place. As one interviewee commented, despite having undertaken a series of retraining courses, recurrent bouts of MIH rendered her unreliable as an employee in the eyes of prospective employers. She held little hope of securing a job, and saw her life as one of a 'permanent student'. Yet as another remarked, those few hours of work available to him through the voluntary sector, were his sole reason for getting out of bed – this employment opportunity thus gave him a sense of worth and dignity – a sentiment echoed by others.

A final issue concerns institutional care and the preferences of those with MIH. Deinstitutionalisation was viewed as the answer to the scandals of ageing and poorly run institutions emerging in the 1960s (Grob 1995); mental health scandals *now* featuring in the media, however, occur in the community rather than in hospital environments. Perhaps in the drive toward the large-scale closure of psychiatric hospitals and the implementation of community-based care, we may be overlooking a key function of such facilities – asylum in its traditional capacity – a place of protection, refuge and recuperation (Wing 1990). Interviewees, for example, spoke of the difficulties of a relapse into MIH when in the community, and the inability of most members of the community to recognise the symptoms. Several spoke of their erratic behaviour, leading to arrest and detention in police cells until their condition was recognised and understood. As Simon commented, 'The police were very good once they realised, but they have no training to deal with it – they're being dumped with a responsibility that is not theirs – even GPs can have difficulties getting patients admitted out of daytime hours.' With few new admissions leading to long-term care, institutionalisation is less of an issue. Interestingly many of those 'revolving door' individuals interviewed commented that during periods of relapse they would prefer to be located within an institutional environment. It was a place where they felt 'safe', allowing them to deal with their problems in an atmosphere of peace and tranquillity. This factor was endorsed by a survey of facility-users undertaken by Dumfries and Galloway Health Council (1993: 3), summed-up by this user's comment:

> I felt safe in hospital. I liked the rehabilitation unit – you could do things
> for yourself, you were not frightened to speak out – it was home-like.

Just as exclusion implies a narrowing of the category of 'us', so the development of a truly inclusive society implies an expanded 'us'. However, whilst the longer term aim must surely be inclusion in the sense of heightened public understanding and acceptance of 'difference' as exhibited by individuals with MIH (and indeed other marginalised groups) this does little to relieve their more immediate needs. It may be that during such periods, the ability of individuals to distance themselves from the stresses of everyday living could be helpful. This does not imply a return to the isolated and exclusionary institutions of earlier decades, but simply that in the provision of small-scale, supported therapeutic 'retreats', we may be able to offer a 'safe haven' within which individuals can focus on their return to mental health. Further, in considering the location of mental healthcare, the voices of those with MIH should be more clearly attended, particularly within rural environments – so far neglected by geographers.

Conclusion

This chapter has focused on three main themes, which, it is suggested, highlight areas that future geographical research might usefully explore. First, it has emphasised a need to move on from a focus on the deinstitutionalisation of those with long-term MIH from institutional environments, to incorporate the 'revolving-door' individuals, whose experiences – due to their retention of greater community ties – may manifest themselves in a different spatial expression of MIH to that demonstrated in earlier studies. Here, it has been suggested that the geographies of the 'new chronic' individual with MIH are influenced by additional sets of social and environmental factors. These stem from the policy and practices of local communities and community-based service providers, the increase in short-term hospitalisation and the needs and desires of those with MIH themselves.

Second, this chapter has focused on issues of exclusion. Whilst institutional care created 'spaces of insanity' in which those with MIH were both spatially and socially excluded from mainstream society, the ethos of deinstitutionalisation and community-based care purports to be 'inclusionary'. As highlighted, however, deinstitutionalised care has created landscapes in which the geographies of those with MIH are shaped by new social and environmental influences and new sets of exclusionary experiences. Whilst earlier studies demonstrated the ability of powerful neighbourhood groups to spatially exclude those with MIH from their communities, this has not been the experience in the Scottish context due to differing legislative and planning practices. To plan or legislate for the cessation of social exclusion, however, which may often be as much perceived as experienced (though no less real to the individual) may be less easily achieved.

Third, the chapter has drawn attention to the paucity of geographical research surrounding issues of rurality and MIH. Despite the changing social and environmental influences impacting on community-based individuals with MIH, most studies continue to focus on urban space. Mental health needs amongst rural populations are poorly understood. Social factors unique to, or exacerbated within rural areas are often inadequately accounted for in the development of community mental health services. Whilst the provision of mental healthcare to rural areas must necessarily be different from that of more densely populated cities, this is not always reflected in the siting of local services. Here, for example, though general medical clinics were found to be geographically well dispersed, statutory sector mental healthcare services displayed a clear urban bias. Whilst the voluntary sector revealed a greater acknowledgement of rural need, and was implementing policies aimed at housing *dispersal*, it has limited resources with which to address the social needs of the population it serves.

The combination of such factors as geographic and social isolation, the perception of heightened visibility and stigma, difficulties inherent in any attempt to develop effective support services to geographically widespread populations, can all contribute towards the creation of 'push factors' impelling those with MIH to seek the anonymity of more populous locales. In Dumfries and Galloway, the social and employment needs of individuals with mental ill-health were shown to be largely un-met outwith core areas of population. Social networks exert strong 'pull-factors' which can override residential location. Here, social isolation and poor employment opportunities appear to have increased travel patterns from rural to more populous areas, as those with MIH find themselves located in dispersed housing alternatives, yet seeking the support of social and employment opportunities more readily available in the urban core. Such a response, however, cannot resolve the social needs of all those with MIH in rural environments. Poor public transport, the high cost of travel for a largely unemployed sector of the population, and the inability of some individuals to access private transport alternatives can create barriers to access.

At a policy level, therefore, it is the contradictions between the locational practices of some agencies operating within the field of MIH, and the needs and desires of service users themselves, which need to be addressed if we are to enhance the community integration of those with MIH. Rural localities in particular face challenges in shaping equitable, yet cost-effective community-based services to geographically widespread populations. Such needs cannot be met by essentially urban-based policies. Unless greater emphasis is placed on decentralised models of mental healthcare, there is likely to be an exacerbation of inequity in the response to MIH, through the effective exclusion of geographically widespread populations from 'community' but urban-based service provision. Whilst the focus here has been on MIH, it should also be noted that many of the issues raised here may be of equal importance to other disabled groups located within rural environments.

Finally, it is acknowledged that this study focuses on only one rural locale whose features cannot be said to be representative of all rural environments. Consequently, it may be that there is a need to develop different typologies of the rural environment (for example, peri-urban, rural, remote) that could help us to develop our understanding of those factors impacting on the spatial geographies of rural-based individuals with MIH.

Notes

1 Bachrach (1989: 165) defines deinstitutionalisation as 'the shunning or avoidance of traditional institutional settings, particularly state mental hospitals, for chronically mentally ill individuals, and the concurrent development of community-based alternatives for the care of this population'.

2 Whilst England and Wales moved towards the integration of psychiatric units within General Hospitals, for example, this has not been the general approach in Scotland. The 1989 report *Mental Hospitals in Focus* advocated instead the development of a 'mental health campus' approach for Scotland (Scottish Home and Health Department 1989: para. 78).

3 A region in south-east Scotland.

4 The term 'chronic' is used here to refer to those who experience recurrent or persistent mental ill-health.

5 Activity Resource Centres' employment projects for those with physical disabilities and learning difficulties.

6 For a more detailed discussion of this attempt at community exclusion see Milligan (1996).

7 Personal interview 19 May 1997.

8 These are defined within the 1989 White Paper as, 'those affected by the problems of ageing and/or physical, mental or sensory disability' (HMSO 1989: para. 2.10).

9 For example, the Mental Illness Specific Grant – ring-fenced funding issued by the Scottish Office.

10 Town and Country Planning (Use Classes) Order 1987, updated in the Town and Country Planning (General Development Procedure) (Scotland) Amendment Order 1993.

11 A voluntary organisation who defines their remit as providing a confidential and supportive befriending service to those going through a period of crisis and in danger of taking their own lives (GCVS 1995).

12 For example, the greater costs, both financially and in terms of time–distance difficulties encountered in delivering services in geographically widespread locales with a low density population base (Milligan 1998).

13 The names of all infomants have been rendered anonymous, in the interests of informant confidentiality.

14 Nithsdale is a district within the Dumfries and Galloway region.

References

Bachrach, L. L. (1989) 'Deinstitutionalisation: a semantic analysis', *Journal of Social Issues* 45, 3: 161–71.

Berry, B. and Davis, A. E. (1978) 'Community health ideology: a problematic model for rural areas', *American Journal of Orthopsychiatry* 48, 4: 673–9.

Blank, M. B., Fox, J. C., Hargrove, D. S. and Turner, J. T. (1995) 'Critical issues in reforming rural mental health service delivery', *Community Mental Health Journal* 31: 511–24.

Bonthron, S. (1995) 'Community care in Scotland: progress from a Community Care Implementation Unit perspective', *Health Bulletin* 53, 6: 356–8.

Burvill, P. W, Stampfer, H. and Reymond, J. (1984) 'Rural psychiatric services in Western Australia', *Social Science and Medicine*, 18, 11: 991–6.

Cloke, P. and Little, J. (1997) *Contested Countryside Cultures*, London, Routledge.

Dean, K. G. (1979) 'The geographic study of psychiatric illness: the case of depressive illness in Plymouth', *Area* 11: 167–71.

Dear, M. and Wolch, J. (1987) *Landscapes of Despair – From Deinstitutionalization to Homelessness,* Cambridge: Polity Press.

Dear, M. J. and Taylor, S. M. (1982) *Not on Our Street,* London: Pion Ltd.

DeLeon, P., Wakefield, M., Schultz, A. J., Williams, J. and Van den Bros, G. R. (1989) 'Unique opportunities for health care delivery and health services research', *American Psychologist* 44, 10: 1298–306.

Dumfries and Galloway Health Council (1993) *Good Care: The Views of Mental Health Users*, Annan: Dumfries and Galloway Health Council.

Dumfries and Galloway Regional Council and Health Board (1996) *Joint Community Care Plan 1995–1998*, Dumfries: Dumfries and Galloway Regional Council.

Eyles, J. (1986) 'Images of care, realities of provision and location: services for the mentally ill in Northampton', *East Midland Geographer* 9: 195–202.

Ferguson, B. and Sanderson, D. (1994) *Review of Adult Mental Health Services and Services for the Elderly Mentally Ill,* York: York Health Economics Consortium.

Foucault, M. (1979) *Madness and Civilization – A History of Insanity in the Age of Reason*, London: Tavistock.

GCVS (1995) *Directory of Glasgow Voluntary Organisations*, Glasgow: GCVS.

Giggs, J. A. (1973) 'The distribution of schizophrenics in Nottingham', *Transactions of the Institute of British Geographers* 59: 55–76.

Giggs, J. A. and Cooper, J. E. (1987) 'Ecological structure and the distribution of schizophrenia and affective psychoses in Nottingham', *British Journal of Psychiatry* 151: 627–33.

Goffman, E. (1961) *Asylums,* New York: Doubleday Anchor.

Grant, M. (1992) 'Planning law and British land use planning system', *Town Planning Review* 63: 3–12.

Grob, G. N. (1995) 'The paradox of deinstitutionalisation', *Society* 32: 51–9.

Hahn, H. (1985) 'Disability policy and the problem of discrimination', *American Behavioural Scientist* 28, 3: 293–318.

Hall, G. B. and Taylor, S. M. (1983) 'A causal model of attitudes towards mental health facilities', *Environment and Planning* 15: 525–42.

HMSO (1989) *Caring for People: Community Care in the Next Decade and Beyond*, Cmnd. 849, London: HMSO.

Human, J. and Wasem, C. (1991) 'Rural mental health in America', *American Psychologist* 46, 3: 232–9.

Hutner, M. and Windle, C. (1991) 'NIMH support of rural mental health', *American Psychologist* 46, 3: 240–3.

Kane, C. F. and Ennis, J. M. (1996) 'Health care reform and rural mental health: severe mental illness', *Community Mental Health Journal* 32, 5: 445–62.

Kearns, R. and Smith, C. (1994) 'Housing, homelessness, and mental health: mapping an agenda for geographical inquiry', *Professional Geographer* 46, 4: 418–24.

Lewis, D. A., Shadish, W. R. and Lurigio, A. J. (1989) 'Policies of inclusion and the mentally ill: long-term care in a new environment', *Journal of Social Issues* 45, 3: 173–86.

McCreadie, R. G. (1982) 'The Nithsdale schizophrenia survey: psychiatric and social handicaps', *British Journal of Psychiatry* 140: 582–6.

McCreadie, R. G., Leese, M., Tilak-Singh, D., Loftus, L., MacEwan, T. and Thornicroft, G. (1997) 'Nithsdale, Nunhead and Norwood: similarities and differences in prevalence of schizophrenia and utilisation of services in rural and urban areas', *British Journal of Psychiatry* 170: 31–6.

McMenamin, F. (1996) 'Castles in the air', *Glasgow On The Line Community Health Magazine*, 10: 22, 30.

MIND (1989) *Mental Health, Housing and Homelessness,* London: MIND Publications.

Milligan, C. (1996) 'Service dependent ghetto formation – a transferable concept?', *Health and Place* 2, 4: 199–211.

Milligan, C. (1998) 'Pathways of dependency: the impact of health and welfare restructuring – the voluntary experience', *Social Science and Medicine* 46: 743–53.

Moon, G. (1988) '"Is there one round here?" Investigating reaction to small-scale mental health hostel provision in Portsmouth, England', in C. J Smith and J. A. Giggs (eds) *Location and Stigma*, London: Unwin Hyman, 203–23.

Murphy, E. (1991) *After the Asylums*, London: Faber & Faber.

Murphy, J. (1992) *British Social Services: The Scottish Dimension*, Edinburgh: Scottish Academic Press.

Murray, J. D. and Keller, P. A. (1991) 'Psychology and rural America: current status and future directions', *American Psychologist* 46, 3: 220–31.

Parr, H. (1997) 'Mental health, public space and the city: questions of individual and collective access', *Environment and Planning D: Society and Space* 15: 435–54.

Philo, C. (1989) 'Enough to drive one mad: the organisation of space in nineteenth-century lunatic asylums', in J. Wolch and M. Dear (eds) *The Power of Geography: How Territory Shapes Social Life*, London: Unwin Hyman.

—— (1992) 'Neglected rural geographies: a review', *Journal of Rural Studies* 8: 193–207.

—— (1997) 'Across the water: reviewing geographical studies of asylums and other mental health facilities', *Health and Place* 3, 2: 73–89.

Prior, L. (1993) *The Social Organization of Mental Illness*, London: Sage.

Pullen, I. (1993) 'Hunting the gowk? – psychiatric community care in Scotland', *British Medical Journal* 306: 710–12.

Robertson, L. J. and McCreadie, R. J. (1991) 'Admission rates, detention rates and socioeconomic deprivation', *British Journal of Psychiatry* 159: 733–4.

Scottish Home and Health Department (1985) *Mental Health in Focus*, Edinburgh: HMSO.

—— (1989) *Mental Hospitals in Focus*, Edinburgh: HMSO.

Scottish Office – Social Work Services Group (1994) 'Community care accommodation: public awareness and local discussion', SWSG Circular no: SW8/1994 ISGO1822.084.

Scull, A. T. (1977) *Decarceration: Community Treatment of the Deviant – A Radical View*, New Jersey: Prentice Hall.

—— (1979) *Museums of Madness: The Social Organization of Insanity in Nineteenth-Century England*, London: Allen Lane.

Shadish, W. R, Lurigio, A. J. and Lewis, D. A. (1989) 'After deinstitutionalization: the present and future of mental health long-term care policy', *Journal of Social Issues*, 45, 3: 1–15.

Sibley, D. (1994) 'The sin of transgression', *Area* 26, 3: 300–3.

Smith, C. J. (1981) 'Urban structure and the development of natural support systems for service-dependent populations', *Professional Geographer* 33: 457–65.

Smith, C. J. and Giggs, J. A. (1988) *Location and Stigma*, London: Unwin Hyman.

Sullivan, G., Jackson, C. A. and Spritzer, K. L. (1996) 'Characteristics and service use of seriously mentally ill persons living in rural areas', *Psychiatric Services* 47, 1: 57–61.

Warner, R. (1989) 'Deinstitutionalisation: how did we get where we are?', *Journal of Social Issues* 45, 3: 17–30.

Wing, G. (1990) 'The functions of asylum', *British Journal of Psychiatry* 157: 822–7.

Wolpert, J. (1976) 'Opening closed spaces', *Annals of the Assosciation of American Geographers* 66, 1: 1–13.

Wolpert, J. and Wolpert, E. (1976) 'The relocation of released mental hospital patients into residential communities', *Policy Sciences* 7: 31–51.

13

ACCOMMODATING DIFFERENCE: SOCIAL JUSTICE, DISABILITY AND THE DESIGN OF AFFORDABLE HOUSING

Flora Gathorne-Hardy

Introduction

The title of this chapter reflects a growing concern amongst geographers with questions of social justice (Smith 1994; Hay 1995; Harvey 1997). This concern has not simply arisen from within the discipline, but is, in part at least, a response to a diverse range of social groups voicing their experiences of powerlessness and social injustice. People with impairments are often at the fore of these political movements, as individuals and disability organisations strive to combat their marginalisation and exclusion from what is presented as 'normal life' (Oliver 1990; Morris 1991). This confluence of political action and critical reflection has revitalised debates about social justice, broadening research agendas to encompass not only more traditional distributional issues but also the socio-cultural, political and economic processes which reproduce distributive injustices (Smith 1994: 7). Counter to oppressive and assimilative ideas such as that of 'normal life', discussion within social and cultural geography has turned to explore human difference and diversity (Women and Geography Study Group 1997). One implication of this current turn is to explore the possibilities of a 'politics of difference' where multiple groups with different needs are empowered to participate in collective, democratic processes of decision-making about issues which affect their lives (Imrie 1996a; Young 1990).

The provision of housing which is designed to meet the needs of physically disabled people is perhaps one of the most difficult areas in which to realise a politics of difference. Housing is a high cost and permanent physical structure. Its production is technically complex and involves a large number of usually able-bodied professionals who are expert in dealing with the regulatory and legal requirements which apply to new developments as well as to rehabilitation and conversion of buildings (Page 1996). In contrast to the permanence of bricks and mortar, people with physical impairments have particular design needs that are likely to change over time. But, few physically disabled people can afford to buy property or pay for necessary design adaptations in the home, which leaves them little choice but to look for some form of subsidised or 'affordable' housing which will have been designed without incorporating their needs or views.[1] Bearing these issues in mind, my aim is to scrutinise current theoretical debates about social justice and decision-making processes surrounding a housing project in the light of each other in order to evaluate the utility of the notion of a 'politics of difference'. The particular housing project that I describe here has been chosen from several that I studied because it was presented by all the main agencies involved as creating an opportunity for physically disabled people to collaborate with others in the design and construction of their own homes. It was initiated in 1994 by a group of housing associations in partnership with a local authority.[2] Five families, each with a physically disabled member in the household, chose to become involved and architects were employed to help translate their different needs and views into detailed architectural plans. It was a self-build development and so members of the five families were expected to carry out the construction work with additional support from volunteers. This case study forms part of a wider body of research I have carried out for my doctoral thesis, which compares the history and current practices of organisations in the UK and the United States which promote ways of enabling marginalised social groups (usually, but not always people on low incomes) to influence the design of their homes.[3]

Just design

'Social justice' is a highly contested concept, being as it is 'a socially constructed abstraction which motivates human conduct' (Smith 1994: 23). There are, however, enduring concerns that course through historical debates. Perhaps the most fundamental of these revolves around the desire to achieve a more fair and equitable distribution of resources amongst members of society. Indeed, there is a vast literature which explores possible combinations of rules and arrangements needed to govern the (re)allocation of material as well as non-material resources in order to achieve distributive justice (Smith 1994). Within recent years this dominant interpretation of social justice has been challenged. Authors such as Iris Young have drawn attention to the limitations of a distributive focus which 'obscures other issues of institutional organisation at the same time that it often assumes particular institutions and practices as given' (Young 1990: 8).

For example, she examines the ways in which welfare state practices which rest upon a distributive paradigm depoliticise the process of public policy formation by defining such decision-making processes as the 'province of experts' and confining conflict to 'bargaining among interest groups about the distribution of social benefits' (ibid.: 10). Thus, she argues that ideas of distributive justice function ideologically, representing the institutional context in which they arise as natural or necessary and forestalling criticism or debate about alternative social arrangements (ibid.: 74). In searching for ways to counter such oppressive practices and 'repoliticise public life', Young emphasises the extent to which diverse social movements including the campaigns of disabled people have given theoretical direction to her work. Her aim, she explains, is 'to express rigorously and reflectively some of the claims about justice and injustice implicit in the politics of these movements, and explore their possible meaning' (ibid.: 7).

What the work of Young and many other contemporary writers point towards is the idea of 'a politics of difference' which rejects assimilationalist ideals and universal norms and instead seeks ways to ensure that socially and culturally differentiated groups are able to participate in collective, democratic processes of decision-making about issues that affect their lives. Although important, an equal distribution of resources is not seen as the ultimate measure of a just society. Instead, attention needs to turn to the ways in which different people experience different forms of domination and oppression and how these processes are reproduced through what may be presented as 'impartial' structures of decision-making, unquestioned cultural norms or physical structures as seemingly innocent as a 'normal' doorway. It is important to recognise, however, that Young's aim is not to deny distributive justice but rather to think through how to *combine* redistribution and recognition. Here her project has parallels with the work of Nancy Fraser and as Ann Philipps, in a recent assessment of both writers' work, has observed that both 'see contemporary societies as characterised by a combination of economic *and* cultural injustices, and neither makes any explicit claim about one being more fundamental than the other' (Philipps 1997: 143).

These commitments to reflect from particular social contexts on notions of social justice and to listen to the concrete social and political experiences of disabled people has informed theoretical debates within geography. On one important level, it has focused attention on the practice of academic research, raising questions about power, the politics of our involvement with people whose lives we seek to represent, and how we imagine our work will influence the thoughts and actions of different audiences (Gleeson 1996; Imrie 1996b). At the same time, while some commentators believe a preoccupation with difference signals to be a worrying shift towards relativism, many other geographers have embraced the move towards more sensitive and complex political imaginaries which attend to the ways in which disability, gender, ethnicity, sexuality, and age are all simultaneously relevant to an understanding of how different people experience place (Butler and Bowlby 1995; Sibley 1995; Duncan and Ley 1993). Given that within the UK (as in many other countries) physically disabled people are experiencing rising material inequality, a fundamental challenge is to explore the possibilities of implementing a politics of difference which integrates socio-cultural differences with claims for redistribution.

In my own research I have sought to engage in a reciprocal evaluation of the theoretical idea of a politics of difference and a particular example of practice, namely decision-making about the design of affordable housing. During this research process, the experiences and views of disabled people have helped me understand the ways in which cultural and economic factors are thoroughly inter-penetrated and intimately bound up with the design of housing. For example, many physically disabled people experience discrimination in terms of employ-ment and so have to rely upon inadequate benefits which fail to cover the extra costs incurred as a result of their impairment. As mentioned above, this means that home ownership is rarely an option. With scarce funding for adaptions, and with anti-discrimination legislation skirting around new residential developments, disabled people are often left living in ill-designed accommodation. Poor design results not only in unnecessary practical problems, but also has an adverse impact on people's dignity, privacy and opportunity for self-determination (Swain *et al.* 1993).

A central question, therefore, is how to achieve a prior political recognition that people with physical impairments have 'different' design needs without creating crude dichotomies between able-bodiedness and physical disability? In my doctoral research, I have explored how to underpin physical planning and housing provision with ideas about diversity, and the paradox of planning for people's changing needs when the physical architecture of buildings is relatively fixed. Disabled people have for many years addressed such paradoxes – what Fraser describes as 'the hard political questions that arise when we try to pursue redistribution and recognition simultaneously' – and acknowledged the need to campaign strategically around different notions of difference (1997: 7). Perhaps the most fundamental debate within the disability literature is one which continues to address the relationship between physical impairment and the social construc-tion of disability. Within this debate commentators have sought to destabilise socially constructed dichotomies between able-bodiedness and disability without displacing people's very real experiences of impairment (Hughes and Paterson 1997). Yet, while disabled people campaign for a recognition of difference and diversity of need, recent legislative changes have not proven promising. The UK Disability Discrimination Bill did not cover residential properties and, as many commentators have noted, the 1995 Act and other legislative changes (such as eligibility for the Job Seekers Allowance) reproduce a negative and inaccurate individual or medical model of disability (Gooding 1996; Chadwick 1996). Opportunities have been missed to refocus policy and action onto 'cultural assumptions, disabling barriers, work practices and organisational structures rather than individuals and their impairments' (Chadwick 1996: 38). Funding arrangements for affordable housing development and the severe cuts in overall funding to the Housing Corporation have also led to worrying trends. Emphasis on cost-effectiveness has resulted in lower space and design standards, while shorter funding time-scales limit the opportunities for consultation with physically disabled people and their organisations about the design of specific housing developments (Allan *et al.* 1996). It seems that policy makers and 'pragmatic' design professionals are still failing to listen to the views of disabled people. Indeed, what is perhaps most notable about affordable housing design in the UK, whether identified as 'special needs' housing or developed without any commitment to promote accessibility or to accommodate the needs of people with physical impairments, is the fact that decision-making continues to be dominated by able-bodied professionals (Knox 1988). As Imrie explains, 'the interplay between the ideologies and institutional practices of the design professions, within the wider context of particular socio-economic structures, has served to exclude minority interests while reinforcing an alienating and oppressive built environment' (Imrie 1996a: 75). Some of the leading architectural journals are beginning to address issues of access. However, the attitude that disabled people's needs are merely one more technical sphere to master and clip on to housing design persists (Clark 1995).

Architecture is just one of the design professions and just one of the many professions involved in the complex process of financing, designing and developing affordable housing within the UK. However, as Imrie suggests, studying the practices of architects provides insights into the effects of power linked with the institutionalisation of 'expertise' in the form of the profession (Johnson 1993: 140). Disabled people have long struggled to uncover these connections between power, knowledge and the geographies of particular places, calling for attention to the ways in which actions of disabled people, their bodies and desires become objects to be disciplined by professionals and governed through the design and management of particular places (Butler and Bowlby 1995). Other authors have explored how the social structuring of professions, such as that of the architecture profession, are both protective and self-reinforcing (Cuff and Ellis 1989). All of these debates challenge the notion of power simply as some 'thing' to be possessed or denied. As Young argues, there are a series of problems with extending a distributive approach to power. First, it tends to obscure the fact that 'power is a relation rather than a thing' and power relations are on-going processes that exist in action (1990: 31). Second, a focus on individuals 'misses the larger structure of agents and actions that mediates between two agents in a power relation'

– structures which can lead to systematic exclusion of social groups from participating in determining their action or conditions of their actions (ibid.: 31). Third,
a distributive understanding of power tends to conceive 'a system of domination
as one in which power, like wealth, is concentrated in the hands of a few' and so
'a redistribution of power is called for which will disperse and decentralise power
so that few individuals or groups no longer have all or most of the power' (ibid.:
32). Instead, Young argues that if we understand power as 'a function of a dynamic
process of interaction within regulated cultural decision making situations, then it
is possible to say that many widely dispersed persons are agents of power without
"having" it, or even being privileged' (ibid.: 33).

So, as Fraser argues, 'justice today requires *both* redistribution *and* recognition . . . [it] also means theorising the ways in which economic disadvantage and
cultural disrespect are currently entwined with and support one another' (1997:
12). This means exploring the cultural and institutional contexts in which decisions about affordable housing design are made and the ways in which notions
of professional expertise marginalise and disempower physically disabled people.
In the next section I want to weave these theoretical ideas into a critical evaluation of a project which was presented by the lead funding agencies as a highly
innovative attempt to enable physically disabled people to collaborate with professionals in the design and construction of their homes.

Ideas of community design

The empirical research which comprises the basis for this chapter should be seen
within a wide context: the history and current practices of organisations which
support low income groups to influence the design of the housing and wider
built environment. People practising within these organisations tend to describe
their work as 'community design' or 'community technical aid'. The core principles of community design are similar to those associated with advocacy planning,
coalescing around the assertion that 'the user knows best' but that the needs
and views of less powerful individuals and social groups, such as physically
disabled people, are systematically ignored by architects, planners, engineers and
other design professionals and so they need assistance in making their claims
audible (Towers 1995). In contrast to the privileging of 'professional expertise',
community designers usually perceive their role as supplying the necessary
resources, training and technical services to ensure that people are 'empowered'
to influence the design of their home environment. Ideas of community design
were first developed by design professionals in the UK during the early 1960s.
Positioning themselves in opposition to top-down urban renewal programmes,
community designers sought to respond to the demands of community groups
and residents who were campaigning for greater involvement in decisions about

their local environments. Non-profit organisations were established, funding sought to provide subsidised technical services to low income groups, and networks of practitioners set up to promote community design and to try to influence policy making. Over the years, discourses of community design and notions of how, when and more precisely why low income groups should partic-ipate in decisions about the design of the built environment have shifted in a rough rhythm with wider economic, political and cultural changes. For example, during the 1980s community designers appealed less frequently to discourses of social justice, moving towards more utilitarian arguments about cost-effective-ness and more euphemistic ideas of 'building community' which resonated with Conservative government ideologies. Therefore, the exact role of community designers as professional advocates seems always to have been ambiguous, while their organisations have been vulnerable due to their dependence on external subsidy. In fact, today there are only a handful of non-profit community design centres still in operation with most community designers working within private practices.

Researching the case study: the background

Before analysing the case study, I want to briefly set out my own position and the methods I used in carrying out research for this chapter. I am an able-bodied woman. My interest in the project stemmed from the fact that it was unique in being a self-build scheme which sought to involve people with physical disabil-ities and one which employed leading community designers to support the residents throughout the design and construction stages. When I first contacted the group in the summer of 1996 the buildings were near completion. Three of the five families had moved in (as all the families had been living elsewhere during the development process) and the resident group had formed a cooper-ative to manage the properties. Having made an initial visit to the site, the residents invited me to stay with them for five days in order to help with the building work and carry out interviews with different members of the cooper-ative. During this time I was able to look at other relevant material (including a TV film made the previous year, detailed drawings and financial accounts) as well as discuss the possibility of writing a short article about the project. I then made two further visits to the site, by which time all the residents had moved in and were able to feed back to me what they felt about the design of their new homes. These visits provided an opportunity to discuss draft copies of mate-rial I had written about the project and to receive feedback from the residents on the way I had presented their experiences and views. I also carried out inter-views with three of the community designers involved, as well as representatives from one of the housing associations. When it came to presenting the material in this chapter, the residents decided it would be more appropriate not to name the organisations involved. I then made the decision to only use the comments from Ann, who is one of the residents, directly in the text as she was the person with whom I carried out most of the detailed interview work and have since worked with in order to publicise the project more widely. All of Ann's comments are quoted from an interview carried out on 30 August 1997.

In the case of this housing project, in 1994 a proposal to develop five homes for families each with a physically disabled member was put forward by a leading proponent of self-build schemes who was working for a London-based housing association. He was able-bodied, as were all the other professionals who became involved in the project. In negotiation with a local authority outside London, a site was identified and a second housing association contacted which was based in the area and through which Housing Corporation funding could be accessed. Self-build proved attractive to all the main agencies because savings on labour costs were estimated to halve the construction budget. Finally, a London-based private architecture practice noted for its community design work was contacted and initial designs drafted for the planning application. It was only at this relatively late stage that the project was advertised in order to find prospective families who could provide the twenty-one hours of volunteer work per person per week – a figure which was to rise throughout the development process.

The experiences of the resident group

All the families who eventually joined the project had been living in housing that did not meet their immediate or long-term needs. For example, Ann has multiple sclerosis and has difficulty walking or carrying out physical movements such as getting in and out of a bath. Ann described the housing situation she was in prior to the project as unsuitable:

> My husband and I were in a situation where we were living in council property at the time, which was extremely small. I had a relapse with multiple sclerosis and was finding it very difficult to get around – my legs were all wobbly and all over the place. The council had tried to put a shower in the bath, but that wasn't satisfactory at all and the doors were extremely narrow – if I was in my wheelchair I couldn't go from one room to the other. Private rented sector was too expensive, we had just moved from that, and of course when you are living in someone else's flat they don't want you to do anything to their property, not even put a bar up, or a hand rail at the side of the loo. As for housing associations, the rents at that time were too high for us and the queues were a mile long.

And then talking to everybody else in the project I discovered that others were living in conditions which were just not suitable. We all had different forms of disability, so therefore all our needs would be different. And so actually getting involved with the architect from the beginning meant that we could each tell him what our needs would be in the long term.

The problem faced by two other women who used wheelchairs was that there were no affordable, accessible properties within the area (including housing association, local authority and private rented accommodation) which included the bedrooms they needed for their teenage children. They felt the assumption was that all wheelchair users were either single, childless or elderly. For a fourth family, their design needs revolved around caring for a severely physically and mentally disabled son, while the fifth family comprised a married couple who were both wheelchair users and who had been living in very cramped accommodation.

Once the group of future residents had come together, they met with all the relevant agencies and began to discuss the development process. First, they formed a cooperative. The initial arrangement was that the London-based housing association would provide relevant training and ensure a steady supply of volunteers, while the local housing association took a backseat position. The financial arrangements were complicated, with money being transferred between the two housing associations and to the community designers, leaving the residents with very little direct control over spending. However, they were very enthusiastic about meeting the community designers and an occupational therapist in order to develop designs for the interiors of the houses. Several key objectives were put forward by the community designers. The first was to find out the specific needs of each the five families; second, to develop flexible design solutions which would accommodate people's needs as they changed over time; and, third, to achieve a high quality of design. Many innovative ideas were discussed including the design of room layouts which created space for wheelchairs, as well as wide entrances and additional storage space for equipment; specific structural changes such as reinforced beams to carry hoists or raising the height of the wiring to make wall plugs accessible; and special design features such as walk-in showers, suitable sink fixtures and lower work surfaces.

Almost immediately it became clear that the disabled residents themselves would not be able to carry out any of the construction work. Efforts were made to identify building tasks, but this was abandoned as it became clear that the building site was inaccessible and there were few tasks suitable for them to undertake. This failure to anticipate such practical problems signalled a lack of understanding of the nature of disability amongst the agencies who had planned the project. It also meant that four able-bodied people ended up carrying out all the building work for their own homes as well as for the fifth couple who were both wheelchair users. Other unhelpful assumptions were also made. In calculating project time-scales, for example, no account was taken of the fact that the members of the families who were carrying out the construction work were also providing care for their disabled relatives. The pressure on the self-builders rose still further when the London-based housing association failed to find volunteers to help with the building work. Inevitably, the project began to run over-time and while the self-builders managed to stay under the planned costs, the overall budget rose because the various agencies involved all continued to invoice the project for their on-going staffing and administrative costs. Such problems were exacerbated by the very poor communication between the different agencies. Ann commented:

> There was no communication, there was a complete lack of commu-
> nication. We would get messages first and second hand. You can't run
> a project like that. You have a site meeting every four to five weeks
> and it is as if [the representatives from the housing associations] are in
> their own little world. And they know how they want it to go, but
> they're not open to any other suggestions.

As the first year passed, it became clear that the same attributes that had initially been seen as highly innovative – that the scheme was to cater for disabled people's design needs and that it was cost-effective – created enormous diffi-culties. In turn, these problems stemmed from the fact that physically disabled people had not been consulted during the early planning stages. As the building progressed the budget came under closer and closer scrutiny. Ann describes the ways in which cost-cutting efforts resulted in changes to the housing design:

> It was 'Oh yes, these properties are going to be for disabled people'.
> But then when they started to realise how much things were going to
> cost for a disabled person, I believe, unfortunately, we got the short
> end of the stick there. A basic thing like a shower tray for a normal
> person – whatever you call normal – for an able-bodied person to go
> and buy a shower tray will spend anything from about £45 to £150.
> You want a disabled person's shower tray, you're looking £400 and
> upwards. Shock horror. Taps, lights, switches, it was the same response.
> When these french windows went in, for instance, I came and had
> a look and I said 'Well, at the moment I'm not in a wheelchair but I
> don't know how long it is going to be before I have to use one all of
> the time. How on earth do we open them, in case of fire?' If you can't
> get out of your wheelchair, look where the lock is. *(Points to top of*

french windows.) We had to put up with that unfortunately, they can't be changed. To be fair, we have got round it. But that's only because the guys [doing the building work] know what they are doing. We can do it this way. We can try and save a bit of money by doing that. It shouldn't be like that. If a property is for a disabled person it should be first and foremost what they need to keep their life independent and to make life easier for them.

What had been presented as an opportunity for physically disabled people to design and build their own homes and what began as a very positive partnership with the community designers became a decision-making process dominated by external funding agencies. Issues of design became marginalised by concern about spending levels in spite of the fact that the construction work was carried out with no labour costs. In 1997, at the end of a long and exhausting process, the residents do feel their design needs have mostly been met. However, this is ultimately due to their extraordinary commitment and strength in confronting professionals who had been either too geographically distant, or who had failed to take the residents' views seriously. As Ann explains:

We all know about the pitfalls of finance and being told what to do by people who don't really know what disabled people are about. They have a nice idea 'Wouldn't this be nice to do this for disabled people'. It's a bit patronising in a way, because we've all got brains and we all know what we want, and it should be channelled with architects, with housing associations. Sit down and listen to what they want, instead of pretending you know what they want when you don't really.

I would have no qualms about doing it again. But *we would know next time*, and that's why I feel quite strongly that people should know what has happened here. Because this would be ideal for other disabled people. It would give them independence. It would keep them out of community care. They could have their carers in their own homes. I think it's brilliant. The concept is brilliant. But then, like all concepts, they sound brilliant in theory – then you put them into practice and it doesn't work.

Next time?

In this final section, I want to return full circle to some of the theoretical issues about social justice and the implementation of a politics of difference. There are a number of points to draw out of this case study. First, the enthusiasm with which all the residents embarked upon the project should not veil the fundamental question as to why self-build appeared to the five households the only way in which they could secure affordable housing designed for their immediate and long-term needs. Other groups who have embarked upon the physically demanding and time-consuming process of self-build, as well as self-build organisations, agree that to sustain motivation and accomplish projects there usually needs to be other driving factors as well as the goal of providing new housing. These might include providing training and employment opportunities for unemployed people, incorporating innovative ecological design features into a home that a private developer would be unlikely to provide, or maybe simply fulfilling a life-long dream to build. None of these incentives applied to the households involved in this project. The fact that the motivating factor was housing *need* signals, I would argue, the extent to which the current systems of affordable housing provision within the UK fail to meet the needs of people with physical disabilities.

A second issue relates to the concept of 'empowerment'. UK housing policy, Housing Corporation guidelines, and the annual reports of many housing association all endorse ideas of community involvement, tenant management, consumer choice and other concepts coalescing around ideas of 'empowering' users and residents to participate in decision-making about the design and management of their homes. What this project illustrates is the need to examine critically some of the political agendas inscribed in such terms as 'empowerment' and address questions of why, when and how physically disabled residents can choose to participate (or not) in such processes. In this case, the residents were invited to become involved in order to feed in their views on housing design at a relatively late stage and carry out the construction work. Ironically, although their involvement in the housing development process was closer (geographically and in terms of the amount of time dedicated) and more physically demanding than for any other party, professionals managed to retain more powerful positions, controlling spending priorities as well as determining the ultimate geographies of people's homes. Explicitly addressing these oppressive processes and implementing a politics of difference requires spending money – a commitment to redistribution absent from many of the participatory initiatives propelled forward as promoting tenant control. Implementing a 'politics of difference' also demands a more sophisticated understanding of power and notions of 'empowerment'. As argued above, power cannot be seen as something that can simply be distributed like money or pieces of information but instead needs to be seen in terms of complex inter-relations and mediated actions. Nor can 'participation' be equated with empowerment when the scope for making choices and decisions is circumscribed by wider institutional arrangements, cultural norms such as ideas of professional expertise or the kinds of divisions of labour found in this project.

This brings me to my third point about the need for recognition *and* redistribution. In the case study I traced some of the inextricable links between processes of decision-making (such as the assumption that the building site would be accessible to all the residents), the politics of resource allocation (such as the amounts of funding allocated to manage the project relative to how much was dedicated to meeting internal design costs) and the physical geographies of the houses produced. I would argue that the fact that some design features were seen by the funding agencies as too expensive or too complex to achieve reflects both an undervaluing of the relevance of housing design in enabling physically disabled people to live independently and, ultimately, an undervaluing of disabled people's lives. On a practical level, the outcome reflects a need for disabled people to be able to gain access to independent sources of advice, technical design services and political support which is sustained *throughout* the development process. The research also highlights the need for disabled people to be involved, *if they choose*, not only in early planning stages but also in processes of project evaluation and long-term management. These conclusions reinforce Ann Philipps's argument that recognising difference and redistributing resources are not alternative ways to achieve greater social justice. Instead, we need to 'build on Nancy Fraser's analysis of strategic choices to ensure that the newer politics around the recognition of difference does not displace an older politics around economic inequality' (Philipps 1997: 153). This case study, I would argue, illustrates some of the potential tensions and dilemmas which underscore these strategic choices. In a time of austerity in terms of state spending levels, there seems to be a very real danger that 'special needs' housing is reinvented as 'innovative' or 'different housing', with these kinds of flagship projects providing shallow reassurance that efforts to 'accommodate difference' are being maintained. Enshrined in depoliticised notions of 'community initiative', such schemes rarely receive funding for rigorous evaluation in order to inform mainstream policy, funding criteria, buildings design standards or design guidelines. Indeed, in this case Ann has been alone in her efforts to find ways to publicise the project and bring to the fore political questions about how to recognise the full diversity of people's housing needs and collectively secure the resources to 'accommodate difference'.

All these issues need to be set within the wider political and institutional context of UK housing policy and systems of provision of affordable housing that meets the needs of physically disabled people. While there are still a number of housing associations seeking to provide for the diverse range of people's design needs, recent trends appear to be eroding rather than improving disabled people's chances of securing a well-designed, secure and affordable home (Kestenbaum 1996). Housing associations appear to have been relieved of any quasi-statutory duty to provide housing for all those living within a locality who are assessed as being in housing need (Cullen 1995). As argued above, there is now a very real danger that competitiveness between categories of 'innovative' housing are being introduced to legitimise overall reductions in state expenditure in the name of 'respect for difference' (Meekosha 1993: 193). For, in the meantime, aggressive cuts in subsidy are being combined with an emphasis on cost reductions and increase in the pace of housing developments (Malpass 1995). One outcome has been a concentration of development within fewer, larger housing associations, leaving smaller cooperative or self-build projects few options but to partner with larger associations in order to access funding. Another has been a shift towards more design and build housing association projects which rely upon a limited selection of blue-print designs few of which are accessible or even suitable for retrofitting with adaptions.

Finally, I concur with Meekosha who argues that 'moving beyond "additive" strategies requires reassessment of concepts of universalism, collective provision, multiculturalism and responsible citizenship' (1993: 190). This is a tall order. However, I believe that this case study contains examples of practice whereby a particular group of people sought to connect their struggle for different housing to longer term goals and alternative visions of a just society within what could be described as the 'politics of location'. The resident group sought to work around ideas that were poorly developed, finding ways to challenge crude categories and eventually succeed in building homes that met their needs. Their experiences have implications for public policy and for the ways in which affordable housing is designed and developed. For example, mechanisms need to be put in place which combat the problems of localism experienced by the residents whereby they felt unable to address the systematic problems which extended beyond their immediate situation. This means that far greater political discretion, financial resources and time must be available to resident groups to enable them to draw upon the services of housing providers, community designers, advocates and other relevant professionals in order to ensure that housing produced is sensitive to their particular needs. Such a strategy has to be accompanied by a commitment to universal entitlement to secure and affordable housing provision as well a politicisation of housing design and what are presented as 'normal' images of the home and home life. Young argues that 'politics, the crucial activity of raising issues and deciding how institutional and social relations should be organised, crucially depends on the existence of spaces and forums to which everyone has access' (1990: 240). I would argue that such 'accessible spaces' need to be urgently created and integrated between all scales of relevant policy formation, including national, regional and local government levels, in order to enable disabled people and their organisations to redirect political priorities radically and devise systems of housing provision which better meet their needs.

Acknowledgements

I would like to thank Ann and all the other residents for their kind help in producing this chapter. I would also like to thank Hester Parr and Linda McDowell for their very welcome comments.

Notes

1 By 'affordable housing', I mean housing for people who cannot afford a free market rent for accommodation and which receives a form of direct or indirect financial assistance from the state, including a subsidy, grant, tax advantage or land donation.
2 The disabled residents I interviewed decided they did not want to name the organisations involved in the project.
3 My doctoral research is funded by the ESRC and is entitled: 'Social justice and the design of affordable housing: a US–UK comparison'.

References

Allan, G., Hudson, J. and Watson, L. (1996) *Moving Obstacles: Housing Choices and Community Care*, Bristol: Policy Press.

Butler, R. and Bowlby, S. (1995) *Disabled Bodies in Public Space*, discussion paper no. 43, University of Reading.

Chadwick, A. (1996) 'Knowledge, power and the disability discrimination Bill', in *Disability and Society* 11, 1: 25–40.

Clark, H. (1995) 'More to life than gadgetry', *HA Weekly* 12 May: 12–13.

Cuff, D. and Ellis, R. (1989) *Architect's People*, New York and Oxford: Oxford University Press.

Cullen, J. (1995) 'The risks of moving in to a grey area', *Inside Housing* 1 December: 12–13.

Duncan, J. and Ley, D. (1993) *Place/Culture/Representation*, London and New York: Routledge.

Fraser, N. (1997) *Justice Interruptus: Critical Reflections on the 'Post-Socialist' Condition*, London and New York: Routledge.

Gleeson, B. (1996) 'A geography for disabled people?', *Transactions of the Institute of British Geographers* 21: 387–96.

Gooding, C. (1996) *Blackstone's Guide to the Disability Discrimination Act 1995*, London: Blackstone Press.

Harvey, D. (1997) *Justice, Nature and the Geography of Difference*, Oxford: Blackwell.

Hay, A. (1995) 'Concepts of equity, fairness and justice in geographical studies', *Transactions of the Institute of British Geographers* 20: 500–8.

Hughes, B. and Paterson, K. (1997) 'The social model of disability and the disappearing body: towards a sociology of impairment', *Disability and Society* 12, 3: 325–40.

Imrie, R. (1996a) *Disability and the City: International Perspectives*, London: Paul Chapman Publishing.

—— (1996b) 'Ableist geographies, disablist space: towards a reconstruction of Golledge's geography and the disabled', *Transactions of the Institute of British Geographers* 21: 397–403.

Johnson, T. (1993) 'Expertise and the state', in M. Gain and T. Johnson (eds) *Foucault's New Domains*, London and New York: Routledge.

Kestenbaum, A. (1996) *Independent Living: A Review*, York: Joseph Rowntree Foundation.

Knox, P. (ed.) (1988) *The Design Professions and the Built Environment*, London: Nichols Publishing Company.

Malpass, P. (1995) 'What future social housing in Britain?', *Housing Review* 44, 1: 4–7.

Meekosha, H. (1993) 'The bodies politic – equality, difference and community practice', in H. Butcher, A. Glan, P. Henderson and J. Smith (eds) *Community and Public Policy*, London: Pluto Press.

Morris, J. (1991) *Pride Against Prejudice: Transforming Attitudes Towards Disability*, London: Women's Press.

Oliver, M. (1990) *The Politics of Disablement*, London: Macmillan Press.

Page, M. (1996) 'Locality, housing production and the local state', *Environment and Planning D: Society and Space*, 14: 181–201.

Philipps, A. (1997) 'From inequality to difference: a severe case of displacement?', *New Left Review* 224: 143–53.

Sibley, D. (1995) *Geographies of Exclusion*, London: Routledge.

Smith, D. (1994) *Geography and Social Justice*, Oxford: Blackwell Publishers.

Swain, J., Finkelstein, V., French, S. and Oilver, M. (eds) (1993) *Disabling Barriers – Enabling Environments*, Milton Keynes: Open University Press.

Towers, G. (1995) *Building Democracy: Community Architecture in the Inner City*, London: University College of London Press.

Women and Geography Study Group (1997) *Feminist Geographies: Explorations in Diversity and Difference*, London: Longman.

Young, I. M. (1990) *Justice and the Politics of Difference*, Princeton, NJ: Princeton University Press.

14

'CAUGHT IN THE CINDERELLA TRAP': NARRATIVES OF DISABLED PARENTS AND YOUNG CARERS

Jane Stables and Fiona Smith

Introduction

In this chapter we wish to examine how the creation of a new care category, 'young carer', has added a new dimension to existing narratives on disability. By drawing on recent media coverage surrounding young carers we will explore how the presentation of these 'children who care' as 'poor innocent victims' and 'brave little soldiers' offers a window through which broader issues can be explored. As such, we will focus on the conceptualisation of childhood, parenting and disability that situates narratives on the lives of disabled parents and their children within a specific social and cultural discourse. Through an examination of television and newspaper coverage that 'raises awareness' by championing the 'plight' of young carers, we will show how the roles and responsibilities within these disabled families are inscribed with various layers of meaning: meanings that create a sense of self and identity that when brought to a level of consciousness can be de-coded to expose the normative notions embedded in them.

By drawing on Barthes (1972) semiological analysis of popular culture we aim to show how the stories of young carers currently being told by the British Press are underpinned by dominant ideological constructions. Whilst research concerned with the ideological construction of disability has focused on exposing the barriers imposed by an understanding of disability as an individualised pathology (Oliver 1990; Barnes 1994; Barton 1989), the media's role in transmitting this ideology has only recently come under closer scrutiny by researchers (see Cumberbatch and Negrine 1992). It is our intention in this chapter to add to these debates by exploring how the media (re)presents and (re)constructs notions of disability in their portrayal of young carers.

The media

The media as a form of communication is both a private and public phenomenon 'through which cultural ideas are produced and reproduced' (Crang 1998: 81). In our discussion of the media we acknowledge that whilst reading and interpretation incorporates individual response, 'meanings' are created as part of a broader process dictated by social and cultural forces. As Bazalgette claims, the media is 'inextricably bound up with the whole complex web of ways in which we share understandings about the world' (1991: 65). Whatever one's position with regard to the power of the media to shape or affect attitudes, the various mass media can be seen as vehicles for transmitting and reinforcing attitudes which operate in the interests of dominant groups (Smith 1980: 254). A pertinant example of attitudes relevant to this chapter is the continued stress on the importance of the institution of the family and, linked with this, strong beliefs about the appropriate roles, characteristics, and activities of parents and children.

Though images of disabled people are becoming more common in soap operas, films and television dramas, they are generally portrayed as victims of their own physical/mental impairment and rarely as competent parents (Cumberbatch and Negrine 1992). Central to our concern in this chapter, however, are the 'non-fiction' texts of the media. In the interplay of figurative and literal images in the texts of newspaper stories, televised news programmes and documentaries, we read a fusion of other people's (re)presentation as a lived experience. According to Smith (1980), the importance of the use of images and characters in news stories has been neglected:

> The unselfconscious admission by other media that they are dealing predominantly with 'fictional' representations contrasts vividly with the assertion of the newspapers that they are dealing with 'fact'. Hence there is little reflection about how 'realistic' characters and situations which feature in news [stories] are – by definition they are presented as factual.
>
> (Smith 1980: 254–5)

News stories and documentaries present their audience with 'a continuous stream of images almost all of which are deeply familiar in structure and form. It uses codes which are related to those by which we perceive reality itself. It appears to be the natural way of seeing the world' (Fiske and Hartley 1978: 17). Thus, we would argue, the role of newspapers and documentary-style programmes in disseminating ideological modes of subjectivity appropriate to disabled people and young carers in society is an extremely important one. The consumption of such images can ignore the production and selection process where the apparent transparency of the text is absorbed without being problematised. As Fiske and Hartley point out, newspaper articles and documentaries are 'human constructs, they are the result of human choice, cultural decisions and social pressures' (1978: 16).

At the time of originally writing this chapter we both had a disabled parent and were confronted by the upsurge in media interest concerning the 'plight' of young carers. Our original paper was born out of a reflexive acknowledgement of our positionality and, as such, at the outset of our work, the media played a powerful role in initially shaping how the 'reality' of the lives of young carers was perceived by us. These media texts had informed and created meanings, constructing identities that positioned us within a discourse. The complexity of this position must be acknowledged as our own experiences of living with a disabled parent did not support the stories being presented. This position and the position of others within this discourse fuelled our interest in the role of the media. Responses to our research project included: 'Oh yes, didn't I see a programme on that? It was so sad, it really upset me' (a friend), and 'I read an article about that in the news recently, [pause] terrible' (a colleague), which emphasised the power of the media to imbue meaning into an arbitrary term that rendered it identifiable to the public at large:

> Programmes thus form a social resource about which people chat, gossip or argue. In this sense shows form social events which can feed into other social occasions rather than be seen as displacing them.
>
> At a second level, TV programmes can also create communities out of people who do not know each other. Some collective identities are based around being an audience or common addressees of a message.
>
> (Crang 1998: 96)

As Morrell argues, 'media forms have a significant part to play in shaping public opinion through what they construct and represent as reality'(1993: 53). Furthermore, initial contact with 'welfare professionals' dealing with young carers heightened awareness of the role that both these professionals and the media adopted as gatekeepers in the selection and portrayal of young carers and their families' experiences. When we began to contact young carers' groups at the outset of the research, welfare professionals were pre-occupied with the fielding of calls from the media demanding stories and specific case studies – case studies specific to the media's own agenda. Notwithstanding the role of these representatives from statutory and voluntary sector groups in controlling access to, and to some extent censoring, or denying individuals a voice in the public domain, this chapter will concentrate on an examination of the media's role in framing, constructing and perpetuating certain 'meanings'; namely, what it means to be a member of a disabled family. As we will show in the following sections, these meanings have an impact on both the way disabled people's lives are conceptualised (denying their heterogeneity) and, ultimately, the way they and their families are dis-abled.

Young carers in the media

Increasing interest in young carers by the media has resulted in a number of documentary films and a substantial amount of national and local press coverage 'exposing' young carers (see Figure 14.1). In telling their stories the mass media have used sensational cases to deliver shocking headlines and scenarios to attract their audience. In addition, charities such as Carers National Association (CNA), Barnardos and Crossroads have used the media to run local and national 'awareness raising' campaigns aimed at both young carers and professionals alike. This coverage by the media of young carers' 'plight' has led to an increase in concern over adequate service provision amongst professionals, as well as creating an 'issue' in the eyes of the public. We begin our examination of the parent/child, child/carer dyad, in the (re)presentation of young carers as an 'issue' in media texts, by drawing on the work of Barthes.

Barthes' (1972) work on popular culture derives from Saussure's theory of semiology: the relationship of the signifier – the written or spoken sign, and the signified in the physical world or world of ideas. In his development of the process of signification (a term used to refer to both the sign and the signified) he situates the construction of meaning within a social and cultural framework. This system, that involves various levels (or orders) of signification, is crucial to an understanding of how meanings are conveyed. For example, the image of a child in the first order of signification (or denotation) is understood as a young girl/boy, however, at a second level of signification (or connotation) a 'child' can mean innocence and/or dependence. In other words, the first level of meaning equates to the literal representation and the second is imbued with social and cultural meanings through which it is interpreted and understood. At the level of connotation, therefore, meanings are inherently subjective and alter over time with changing attitudes, values and emotions.

At the level of denotation, the term 'carer', when used to describe a person who cares for a disabled relative, is implicitly accepted by society to mean an able-bodied adult woman. At the level of connotation attitudes of love, duty, sacrifice, and pity combine with this interpretation to create a socially and culturally situated meaning that draws on notions of disabled people's dependency. When we consider the term 'young carer', the juxtaposition of 'child' and 'carer' creates an oxymoron, thereby defamiliarising social and cultural constructions of their individual meanings: confronting and contradicting our common understanding. The idea of a young (child) carer sets up a series of contradictions – an adult is dependent on a child/children are dependent on adults – that combine to threaten ideological constructions of both childhood and parenting. As Crang states:

> Identity can be defined as much by what we are not as by who we are.
> This is where geography often comes in since these 'us' and 'them'

groups are often territorially delimited . . . It will be suggested that space is crucially involved in defining 'other' groups. A process often termed Othering through which identities are set up in an unequal relationship.

(Crang 1998: 61)

As we will now explore, the issue of young carers/disabled parents has centred its concern within the private sphere of the family, providing contested notions of domestic spaces as places where children are cared for by adults. For geographers, 'home' can be understood to mean a literal space: 'a physical reality' as well as a notion that is socially and culturally produced (WGSG 1997: 5–7). This second, and more recent understanding of space draws on Barthes' connotative level of understanding, whereby, the idea of 'home' as a private, domestic space is infused with emotional, social and cultural meanings.

The fluidity of meaning attached to these terms are therefore implicitly situated within a dominant discourse, which we shall now explore by examining one of the many articles that has recently appeared in the British Press.

Victims of the Cinderella trap

Children, as subjects, make emotive news stories. Young carers are eminently newsworthy, offering, as one child welfare professional reflected, 'good photo opportunities'. The messages delivered through such attention grabbing headlines and captions as 'Trapped: Disabled mother relies on son', 'Labour of Love', 'Brave: But mother fears for son' (*Mail on Sunday* 30 March 1997), 'Prison Sentence' (*The Guardian* 2 March 1995) are of the tragic child, the heroic child, and the helpless, dependent, self-centred disabled adult. Whether these young carers are portrayed as saints or slaves or, as one headline claimed, victims of 'The Cinderella trap' (Community Care Conference Pack 1995), the scripts, predominantly relayed through selected case studies, draw on the victim/villain allegory – a tried and tested plot, whose codes and conventions are recognised and, as such, it draws its audience. As one young carer, who has been regularly approached to take part in television and press interviews, claimed, '[t]hey want a sensational story with lurid tales of a horrific childhood'. Reflecting on his own experience of life he adds, 'But it isn't like that and it seems they don't want to accept this. When I tell them what our lives are really like, that they can be rewarding, they are just not interested' (Community Care Conference Pack 1995). Thus it seems that cases selected by the media paint only a partial picture of young carers' experiences. Whilst researchers working with young carers have shown that the lives of young carers vary from one family to the next (Aldridge and Becker 1993; Dearden and Becker 1995; Frank 1995), the media's collective representation of young carers' 'heroism' denies this heterogeneity.

We only have the space to provide one example of how young carers are represented in the media. However, as the media have shown, examples of personal and individual family stories are a powerful tool for engaging an audience. Their selection, however, as has already been suggested, also affects the way certain messages are delivered to an audience. The content of these messages will be made more explicit in the discussion that follows. The newspaper example we are using has been selected and edited by us to examine and illustrate these points. This article appeared in *The Independent* on 3 August 1995 (see also Figure 14.1).

NINE YEARS OLD AND LOOKING AFTER SICK MUM

Looking back on the 12 years since her mother was diagnosed as having multiple sclerosis, Jenny . . . admits her main emotion was anger as she coped with a difficult and frightening situation.

Her mother, Heather, who was divorced, became ill when Jenny was seven. She and her nine-year-old brother, Paul, cared for their mother for two years until he left for boarding school. After that Jenny, then nine, coped by herself. 'I would lift mum in and out of bed, empty the bedpans, look after her generally.'

Her response was human: 'I was angry with mum, I was angry with her disease and I just didn't want to talk about it to anyone. It was very difficult for both of us. Our relationship was quite strained at points. I was often coping very badly as a carer.'

Her greatest problem was that no one listened. 'No one ever asked me what my needs were. Everyone always went on about mum and how she was and as a child I wanted to say what about me?'

Compared to other young carers, Jenny says she was lucky. She did not miss any school and there was regular help once a week . . .

Jenny stresses she does not regret the 11 years she spent caring for her mother: 'I do love my mum and I wanted to care for her. But it's difficult. People have got to remember that young carers aren't "little angels". Each young carer should be treated as an individual who needs support themselves.'

Whilst we acknowledge that this story may not be representative of all young carers' experiences, it is to some extent representative of the media's portrayal. Although the codes and conventions used may alter between newspapers and vary from one medium and genre to another, the underlying messages contained within their stories are underpinned by normative notions of childhood, parenting and disability. These normative notions will now be explored by drawing on the media example above to illustrate how the construction of young carers' lives

261

Child carers need help themselves

Social services departments have little idea what is happening to the thousands of young people who care for sick and disabled relatives, according to a new survey by *Community Care* magazine.

Launching its campaign "Young Carers — Back Them Up", the magazine revealed that almost half the departments have no staff dedicated to young carers and more than three-quarters have no special training.

Young carers, of whom there are an estimated 40,000 in Britain, are defined as people under 18 who care for a parent, sibling or close relative. They deal with a variety of illnesses — physical and mental, alcoholism, drug abuse, Aids.

Research into the ages, gender and types of illness of those young carers cope with will be completed later this year, but preliminary research in Nottingham suggests that children as young as five care for parents and the majority are under 10.

The survey, of 60 directors of social services in England and Wales, also found that 35 per cent will not be able to provide the assessment of young people's needs that the new Carers (Recognition and Services) Act will require. A further 35 per cent did not know whether their budget would cover these services or not. The 6.8 million carers of all ages are estimated to save the Government £30bn a year.

More than three-quarters of departments had no current budget for services specifically aimed at young people and half did not provide practical help for young carers, with 23 per cent having no plans to introduce it in the future. In spite of this, 55 per cent defined young carers as "children in need" under the Children Act.

Professions most likely to meet young carers had little contact with social services in this field: 67 per cent of departments had no formal contact about young carers with educational professionals, 54 per cent did not liaise with GPs and 41 per cent did not liaise with other National Health Service agencies.

Terry Philpot, editor of *Community Care*, said: "Young carers need to have more than applause — they need help based on respect for themselves and their families. Social services departments should not assume these young people can cope. They must urgently start talking to other professions about young carers' needs."

Figure 14.1 Child carers need help themselves
Source: by Glenda Cooper, *Independent*, 3 August 1995

reinforces a dominant discourse on disability; one which is grounded in notions of disabled people's dependency and asexuality.

We are children, not carers

Predominant cultural constructions of, or conventions associated with, childhood are mythologised at a connotative level in Jenny's story. As Fiske and Hartley claim 'the myth by which we apprehend childhood as a free, unconstrained, [and] happy experience' (1978: 57) is of significance here in that young carers are signified by what they are not – they are not carefree, or according to this article, cared for (a point that will be taken up later). Jenny, the former young carer in this article (who, incidentally, was expected to care and attend school whilst her brother was sent away to boarding school), is reflecting on her experience of childhood.

Identifying herself now as a young carer, she is measuring herself on some image of what a carer should be and whatever that image is she does not live up to her understanding of it, as in her words, she 'was often coping very badly as a carer'. As an adult reflecting on notions of young carers, Jenny now situates herself within an adult discourse. Often children are unable to recognise themselves in adult depictions and constructions of young carers' lives. In the limited amount of research that has given young carers a voice, children have not identified with the 'given' representations of their lives. According to Morris (1995), children see themselves as children not carers!

The home: site of confusing emotions

The use of the word 'human', in the extract, to describe Jenny's response of anger towards her mother interpellates us as readers to accept her reaction as natural and therefore not question it. However, Aldridge and Becker's (1993) work with young carers has suggested that children's angry responses are a manifestation of a mixture of anxiety, confusion, fear and isolation. Furthermore, these emotions are based less on some unqualified notion of private family emotion, developed within the confines of the home, but instead, are a reaction to the lack of public acknowledgement, collective support and provision of services to disabled families.

In the case of some young carers, the idea of home as a safe and secure place is contradicted by their expressions of anxiety, confinement and responsibility. In the reporting of young carers in the media these expressions of fear, and the restrictions that caring place on the 'normal' social spaces of childhood, are exemplified:

> Peter's (young carer) biggest worry is his mother . . . 'When Mum first became ill, it was scary,' . . . 'It's sad that I've not formed any strong friendships because I can't go out much'.
>
> (*Mail on Sunday* 30 March 1997)

Moreover, the significance of spatial boundaries is repeated in a study of fifteen young carers where Becker and Aldridge claim:

> If the young carers did go out, they were often the victims of what we have termed the 'caring curfew' . . . if they went out they had to be home early.
>
> (1995: 47)

However, this issue of the spatial control of children in public space is by no means unique to young carers, as Valentine (1995) and Smith (1996) have shown in relation to children in general. The current upsurge in studies of children's geographies reveal that all groups of children are being increasingly confined to the private/domestic sphere:

> There has been relatively little account of the ways in which domestic space is regulated for children . . . This lack in itself speaks to the privatising of childhood in domestic space and the insulating role which the home takes on for children.
>
> (James *et al.* 1998: 54)

Conflict of interests

Although the feelings expressed by young carers in relation to their 'home' life are very real, they are framed within a context that evokes images of their parents as dependent, helpless, demanding, disabled individuals. The audience's understanding of both the disabled parent and the young carer(s) derive from the personal tragedy theory of disability. In these descriptions of carers' lives, the social processes which combine to restrict, not only the children, but also the parents themselves, are ignored. Becker and Aldridge (1995) also pointed out that, 'The majority of the young carers interviewed experienced economic deprivation', a factor that they claimed exacerbated the stress of caring.

Furthermore, in our extract, no one asked Jenny when she was a child what her needs were. No one listened to her, thus denying her a voice, adding to her sense of isolation and fear. Rather than simply being a normal human reaction her feelings are, in part at least, a consequence of society's neglect of disabled families. They are, however, also a consequence of society's failure to listen to children and young people in general as Smith and Barker (1999) and Matthews and Limb (1998) have highlighted in their recent discussions of the lack of children's participation in policy-making in the United Kingdom.

We have already suggested that these stories depict the children as voiceless victims. As such, these news stories demand a villain, someone to blame. Frost and Stein's (1989) articulation of stigmas associated with child welfare problems, although not specifically aimed at young carers, seems to provide an opportune reflection on the dilemmas of apportioning blame that predominate in debates surrounding young carers:

> the 'inadequate' parent, the 'problem' family, the 'incompetent' social worker or the 'bureaucratic' welfare agency are all blamed.
>
> (1989: 1)

Furthermore, it has been asserted that in 'raising awareness' of their needs and the problems young carers experience as children, attention has been skewed away from the underlying problems associated with the provision of services to meet the needs of disabled parents, who, to some extent, are being constructed by the media as Frost's 'inadequate' parents, in one of societies 'problem families'. 'Conflict of interests', an article published in *Community Care* as part of their young carers' campaign in 1995 stated that, 'Some activists feel the focus on young carers disempowers and detracts from the needs of disabled parents' (Cohen 1995: 14). Moreover, in concentrating on young carers, Parker asks, are we in danger of 'ignoring those environmental, social and economic factors which discriminate against disabled people and make it more difficult for them to continue their parental activity?' (cited in Cohen 1995: 14).

Pathologised discourses of motherhood

The system of meaning by which we come to recognise Heather, the disabled parent in the article (dependent mother), is explained through the conflation of denoted and connoted notions of motherhood. Implicit and explicit in the framing of this news report is the presentation of the mother as a difficult and frightening situation (rather than the situation itself). In part the article implies that the blame lies with the mother and her illness. Furthermore, according to this newspaper article, the mother now had to be coped with, cared for and 'looked after generally'. This is reflective of both notions of what constitutes a 'proper' mother (Smith 1996) and society's perceptions of what disabled people are capable of. Not only was Heather a single parent, but she was also now a parent unable to cope, in other words, an inadequate parent. Normative expectations of the motherhood/childhood dyad define children as dependent upon their mothers for nurturance and care. As Pinch and Storey (1992) and Gregson and Lowe (1994) point out, adult women remain responsible for the processes of social reproduction. The language of this article clearly challenges this notion when it positions the mother as dependent on her daughter, Jenny, for care. The media, which as Dyck (1990) argues, is part of the professional discourse which constructs mothering, has pathologised some mothers through stories such as that of Heather.

Therefore, in this representation, what altered was not any dramatic role reversal between mother and daughter but the fact that the mother was no longer able to do a number of tasks herself (e.g., getting out of bed and toileting herself) and, with limited or inappropriate support, her daughter had to carry out those tasks. As Hicks points out, 'another problem the long-term sick and dependent may have to come to terms with is being a burden on someone else and seriously affecting the quality of his or her life' (1988: 147). Lonsdale (1990) describes the concern mothers in her survey felt towards the additional responsibilities their children had to cope with. The resulting relationship has been described by Shearer (1981) as 'mutually handicapped'. The disabled dependent may be unable to ask for the help they need, their carer unable to offer it outside a climate of general uneasiness, and both may be reflected in Jenny's sense of anger and strain within the relationship with her mother.

The contested constructions of childhood and parenting offered in this narrative reveal something of the way disability is conceptualised in society. Disabled men and women are not expected to be parents. It has been suggested that the image of being an inadequate parent is common for women who are disabled. Such women, Lonsdale (1990) argues, are actually 'actively discouraged' from the mothering role which adds to their stigmatisation by society. The ideology that underpins norma- tive notions of motherhood helps construct acceptable images of mothering and cir- cumscribes defined norms of behaviour (Smith 1996). It is assumed that disabled women do not and cannot fit into the prescribed role. Physical disability represents a threat to the expectations we have of what makes a 'good' and a 'proper' mother:

> This is due to a number of things: a belief that disabled women would not be able to cope, that they are asexual, that the children of disabled parents will suffer, that the disability can be inherited, and that rearing requires physical mobility and dexterity.
>
> (Lonsdale 1990: 76)

It could be argued that what is being suggested in this article is that with the onset of disability, or illness, Heather's role as parent and her responsibilities as a parent were relinquished and that the child not only took on the day-to-day responsibilities of performing intimate tasks for her mother but she also took on the role of parenting the parent. However, the concept of parenting is more complex than the ability to complete a number of domestic tasks:

> There is a clear assumption . . . that to receive help with any of the daily activities of going to the toilet, having a shower or taking medicines is to relinquish control . . . Moreover, much of the thinking around the issue of children who are 'carers' has difficulty distinguishing between parent- ing – the concern and sense of responsibility that parents have for their children's welfare in all its manifestations – and the practical and physical things which adults do when looking after children and running a home.
>
> (Morris, cited in Social Services Inspectorate's Report 1995: 42–3)

Moreover, while Parker (1994) claims that young carers may find that they are taking on more responsibilities than their peers, this is not simply a response to their parents' disability but is structurally bound up in the amount of support they receive. Within this discussion there is an implication that the burden of these added domestic responsibilities destablises and confronts normative notions of childhood as carefree. However, Solberg's (1997) work with Norwegian families has revealed that children in general undertake a number of domestic tasks and that given the choice children actually prefer to be given full responsibility at certain times for the domestic work they carry out.

The site of this responsibility is, in itself, contested. Solberg (1997) confronts normative notions of ownership and responsibility surrounding domestic space, fundamentally perceived as an adult woman's space. The separation of public and private space is exposed here in that there is a tension between the activities that a woman and/or mother carries out as a part of everyday life (not usually considered as work) and the requirement of children to participate in what then becomes reclassified as work in their cases. Furthermore, in carrying out this 'work', their re-defined role as child workers poses a threat to their childhood status:

> space is always open to contestation by different individuals or groups, many of whom are trying to question and redefine the meanings and boundaries of particular spaces.
>
> (WGSG 1997: 7)

Conclusion

In this chapter we have argued that the way individual and collective identities are created is informed by the ideology that maintains dominant discourses. We have suggested that in its depiction of young carers, a very narrow range of ideologically distorted stereotypical images of disabled people is employed by the various mass media, serving to reinforce an ideology of personal tragedy, and dependence. By drawing on Barthes' theory of semiology we have suggested that the way the media represents young carers and disabled parents draws on notions of relational identity: where they are defined by what they are not. This Cinderella tale is yet another fictional stereotyping of disabled people, adding to their marginalisation in society.

References

Aldridge, J. and Becker, S. (1993) *Children Who Care – Inside the World of Young Carers*, Loughborough: Loughborough University.

Barnes, C. (1994) *Disability and Discrimination in Britain* (2nd Edition), London: Hurst and Co.

Barthes, R. (1972) *Mythologies*, London: Vintage.

Barton, L. (1989) *Disability and Dependency*, London: Falmer Press.

Bazalgette, C. (1991) *Media Education*, London: Hodder and Stoughton.

Becker, S. and Aldridge, J. (1995) *Young Carers in Britain*, London: Loughborough University.

Cohen, P. (1995) 'Conflict of interests', *Community Care* 24–30 August: 14–15.

Community Care Conference Pack (November 1995) Young Carer's Conference, London.

Crang, M. (1998) *Cultural Geography*, London: Routledge.

Cumberbatch, G. and Negrine, R. (1992) *Images of Disability on Television*, London: Routledge.

Dearden, C. and Becker, S. (1995) *Young Carers: The Facts*, Loughborough: Loughborough University.

Dyck, I. (1990) 'Space, time and renegotiating motherhood: an exploration of the domestic workplace', *Environment and Planning D: Society and Space* 8: 459–83.

Fiske, J. and Hartley, J. (1978) *Reading Television*, London: Methuen.

Frank, J. (1995) *Couldn't Care More: A Study of Young Carers and Their Needs*, London: The Children's Society.

Frost, N. and Stein, M. (1989) *The Politics of Child Welfare: Inequality, Power and Change*, Hemel Hempstead: Harvester Wheatsheaf.

Gregson, N. and Lowe, M. (1994) *Servicing the Middle Classes: Class, Gender and Waged Domestic Labour in Contemporary Britain*, London: Routledge.

Hicks, C. (1988) *Who Cares: Looking After People at Home*, London: Virago Press.

James, A., Jenks, C. and Prout, A. (1998) *Theorizing Childhood*, Cambridge: Polity Press.

Lonsdale, S. (1990) *Women and Disability: The Experience of Physical Disability Among Women*, London: Macmillan.

Matthews, H. and Limb, M. (1998) 'The right to say: the development of youth councils/forums within the U.K.', *Area* 30, 1: 66–78.

Morrell, C. (1993) *Media Education: Representation and Reality – Part 2: Media Practice*, Exeter: University of Exeter.

Morris, J. (1995) 'Creating a space for absent voices: disabled women's experience of receiving assistance with daily living activities', *Feminist Review* 51 (Autumn): 68–93.

Oliver, M. (1990) *The Politics of Disablement*, London: Macmillan.

Parker, G. (1994) *Where Next for Research on Carers?*, Leicester University: Nuffield Community Care Studies Unit.

Pinch, S. and Storey, A. (1992) 'Who does what where? A household survey of the division of domestic labour in Southampton', *Area* 24: 5–12.

Shearer, A. (1981) *Disability: Whose Handicap?*, Oxford: Basil Blackwell.

Smith, F. (1996) 'The geography of out of school childcare', unpublished Ph.D. thesis, Reading University.

Smith, F. and Barker, J. (1999) 'Learning to listen: involving children in the development of out of school care', *Youth and Policy*, forthcoming.

Smith, R. (1980) 'Disabled images', cited in Ellis, J. (1992) *Visible Fictions: Cinema, Television, Video*, London: Routledge.

Social Services Inspectorate (1995) 'Young carers: something to think about', papers presented at four SSI workshops, May–July, Department of Health.

Solberg, A. (1997) 'Negotiating childhood: changing constructions of age for Norwegian children', in A. James and A. Prout (eds) *Constructing and Reconstructing Childhood*, London: Falmer Press.

Valentine, G. (1995) 'Stranger-danger: the impact of parental fears on children's use of space', paper presented to Building Identities: Gender Perspectives on Children and Urban Space International Conference, Amsterdam.

Women and Geography Study Group (WGSG) (1997) *Feminist Geographies: Explorations in Diversity and Difference*, Harlow: Longman.

15

BODY POLITICS: DISABLED WOMEN'S ACTIVISM IN CANADA AND BEYOND

Vera Chouinard

Introduction: missing in action?

When you think of political action by women, what sort of images come to mind? Perhaps you imagine suffragettes marching for the vote, women making impassioned speeches at a podium, or women gathered in front of institutions like courts or government to protest how issues such as male violence against females are handled. If, like many feminists today, you are sensitive to differences amongst women, then you probably imagine that these women are a diverse group: some young, some old, some who have brought along young children, some white, some black, some Asian and maybe some women whose dress or manner signal a sexuality that is non-heterosexual. Chances are, however, that none of these women have a physical or mental disability.

Why are disabled women often missing from our images and accounts of political action by women? In part this is the result of the meanings attached to disabled bodies in the culture of late capitalist societies. The term 'dis-abled' conveys the message that these bodies are negatively different from the able-bodied ideal: less able to take part in normal activities and, more importantly, to succeed in socially valued ways. Such cultural meanings have important social consequences, such as contributing to devaluation in and exclusion from workplaces, and making it more difficult for disabled people to challenge negative readings of their bodies and potential. For disabled women, cultural inscription of the disabled body as negatively 'other' is compounded and exacerbated by gender relations of power. Disabled women are the 'other others' of the disability and women's movements; marginalised by male power within what some have called the 'last civil rights movement', they have also had to contend with indifference and oppression within a women's movement in which the able body norm is seldom questioned.

In this chapter, I examine how disabled women have contested these barriers to activism through struggles for political voice and influence at national and international scales. The chapter begins with a discussion of disabled women's experiences of political activism in Canada.

Here I outline major developments in disabled women's struggles for political voice and examine the kinds of barriers to a body politics of empowerment that disabled women have encountered within both the disability rights and the women's movements. State restructuring is playing a critical role in exacerbating barriers to disabled women's activism as the next section illustrates with recent examples from Canada. Still, disabled women continue to fight back against seemingly overwhelming odds: developing new initiatives, alliances and networks at the global as well as local and national scales. In the final section of this chapter I discuss some of these international organising efforts and the challenges they face. The chapter concludes with reflections on how disabled women can build on these struggles to create an inclusive and empowering body politics, and on the possible benefits of conceptualising disabled women's activism as part of a general politics of bodily diversity, as well as a politics of body difference.

Before proceeding, I would like to add a personal note and reflection. When the editors for this book reviewed the first draft of this chapter, one of their comments was that the account was rather 'depressing'. I could see their point. After all, who likes to hear that disabled women are among the poorest and most ostracised members of society? However, when I thought about how this might be remedied, I concluded that I had no choice but to 'tell it like it is'. Yes, the circumstances in which disabled women live and struggle is depressing. Yes, the fact that social programmes to assist them are being cut or undermined is terrible. But, at the end of the day, this is the reality of disabled women's lives and it is a reality which throws into relief just how hard we have to struggle for social change. And it is the reality of being survivors and refusing to be silenced, whatever the odds, that is the uplifting part of disabled women's stories. That is what we hang on to, and that is how we find the strength to make a difference. This is an account of the differences disabled women are making; with no apologies for the depressing bits.

Making marginalised bodies visible: disabled women's struggles in Canada

If struggles for political voice and presence by disadvantaged groups like women are regarded as occurring at the margins of power, then disabled women's struggles have been launched from the 'margins of the margins'. Economically, disabled women are amongst the poorest members of Canadian society. In 1991, disabled women aged 15 to 34 earned an average of 68.7 per cent of the employment income earned by disabled men in the same age range (who in turn received an average of only 70 per cent of the earnings of non-disabled workers in the same age range). For older disabled women, aged 35–54 and 55–64 average employ- ment income was only 54.7 per cent and 62.6 per cent respectively of that earned by disabled men in the same age ranges. Unemployment rates for persons with disabilities are high: in 1991, Canadians with disabilities aged 35–54 and disabled women aged 55–64 were almost twice as likely to be unemployed as non-disabled Canadians (Statistics Canada 1995). In 1997, the national unemployment rate for women with disabilities was reported at 70 per cent (DAWN Ontario 1997). Poverty is an inescapable fact of life for most disabled women. Barile (1993) char- acterises disabled women in Canada as an exploited underclass: with 65 per cent living in poverty (as compared to 43 per cent of disabled men), many denied pub- lic benefits on the basis of household status (most disabled women live alone), receiving especially low wages (particular in sheltered workshops), lacking union protection if employed, and denied access to many community services (e.g., women's shelters).

Politically, disabled women have found themselves at the margins of the disabil- ity rights and consumer movements as a result of male privilege and power, and at the margins of the women's movement as a result of ableism. In short, their loca- tions within bodies which are not only female but also disabled has intensified and compounded their oppression, and has meant that spaces of resistance for women and for men with disabilities have been spaces of oppression for disabled women.

Disabled women have, however, engaged in significant challenges to their economic and political marginalisation. In what follows, I outline some key devel- opments in disabled women's activism in Canada, and assess the impacts of these on disabled women's capacities to struggle for social change.

Disabled women's activism in Canada

One of the consequences of living at the margins of the margins of social power is that much of the story of disabled women's activism in Canada remains untold. The existing literature documents some important developments but contains significant gaps as well; for example, relatively little information is available on disabled women's experiences of struggles within the disability rights and consumer movements. Nor is a great deal written about political activism itself; in part, perhaps, due to the widespread perception of persons with disabilities as passive and dependent. The following account is, then, a partial one.

For some, the emergence of the Disabled Women's Network (DAWN) at national, regional and local scales marks the beginning of the disabled women's movement in Canada. DAWN developed at the margins of both the women's and disability movements; and the tensions of that position continue to shape the pol- itics and impacts of the organisation and its chapters. The impetus for DAWN came from a meeting between representatives of Secretary of State for Women's Program and four disabled women activists to discuss the possibility of a national conference on disabled women's issues. Interestingly, this meeting was initiated by the state. The period from 1983 to 1985 also saw efforts to increase the visibility of disabled women's issues within the women's and disability movements. This included liaison with the National Action Committee on the Status of Women (NACSW) and pressure within the Coalition of Provincial Organizations of the Handicapped (COPOH) for a national conference workshop on women's issues and an investigation of women's role within the organisation's decision-making structure (Stone and Doucette 1988; Stone 1989; Doucette 1991). One of the women involved in these initiatives was also involved in organising the national conference on disabled women's issues. Whether such developments signalled widespread grass roots activism amongst disabled women in different regions and in other women's and disability groups remains, unfortunately, unclear.

The meeting between disabled women activists and state representatives led to a June 1985 national conference on disabled women's issues in Ottawa. Attendance was small, with seventeen delegates present. Efforts were made to select delegates on the basis of leadership ability, minority status, regional representation, and cross or different disabilities (Stone 1989). However, it is not clear whether the four conference organisers controlled the selection process and if not, how widely other activists or groups were consulted. Nor is it evident how representative the delegates were since published accounts fail to indicate, for example, the types of disabilities and minorities represented at the conference (e.g., Doucette 1991). It was at this conference that DAWN was formed and a committee of eight women established to coordinate its activities. Significantly, membership in the core committee was restricted to delegates with access to computers for communication purposes.

By late 1986, chapters of DAWN had been established in PEI, Ottawa, Halifax, Montreal, and provincially in British Columbia and Ontario. However, the organisation was hampered by a lack of funding, difficulties in maintaining communications, and lack of a clear internal decision-making structure. There were also reports that too few women were shouldering the work of the organisation, and that this was contributing to problems of ill health, burn-out, and disillusionment amongst members (Stone 1989). In March 1987, a second, federally funded national conference in Winnipeg was attended by twenty delegates. As at the founding conference, delegates agreed to promote local membership in the organisation but, once again, efforts were hampered by lack of funds. By March 1993, the national DAWN office, which had served as a link to federal government agencies and international organizations, was forced to close as a result of federal funding cutbacks. With federal funding cut by more than half, DAWN lacked sufficient funds to even hold a national board meeting. At an Ottawa press conference, DAWN representative Pat Israel summarised the organisation's reaction to further funding cuts: 'We refuse to die a slow death as they cut back more and more each year. We'd rather die quickly right now' (cited in Briscoe 1993). Joan Meister, past chair of DAWN Canada, argued that the government was prepared to profit from DAWN (e.g., through positive press) but would not give the funding needed to run the organisation. She noted that DAWN Canada was forced to operate from her bedroom office for many years due to lack of funds (Briscoe 1993). At the time of writing, DAWN Canada continues to operate but its influence and activities appear limited to specific projects such as a recent research report on disabled women and smoking cessation, and organising the recent national conference in Vancouver 1997. Funding cuts have also impacted regional and local DAWN chapters, for example, forcing DAWN Toronto to close in 1996 and DAWN Ontario to move from Toronto to Sudbury in northern Ontario, in order to reduce costs (DAWN Ontario newsletters, February and July 1997).

DAWN's political aims have always been ambitious ones. At the 1985 national founding conference the key aims of the organisation, as summarised by DAWN Toronto, were:

i) [to] make women's services and women's movement accessible to all disabled women ii) [to] act as a bridge between disabled consumer movement and women's movement iii) [to provide] role models for disabled girls iv) [to] do outreach to all disabled women, including native, black, Asian and other women of colour, immigrant women, lesbians, women in institutions and single parents vi) [to] work in coalition with others concerned with social justice vii) [to] provide information on disabled women viii) [to] be the voice of disabled women.

(Stone 1989: 131–2)

More recently, DAWN Ontario has summarised its goals in the following way:

keep current on issues facing women with disabilities; provide role models for girls with disabilities; develop resources for girls with disabilities; help start and support DAWN groups across Ontario; speak for the rights of women with disabilities to make sure we can take part in women's groups, activities, events and services; work with other women's and disability groups; produce resources about health care for women with disabilities; lobby the government on issues affecting women with disabilities. Issues such as employment, advocacy, training, education, transportation, housing, health care, and others.

(1997: homepage, 2)

As I indicate in the next section, disabled women involved in DAWN have faced numerous socio-spatial barriers to effective political voice and activism.

273

Some of these, such as the need to subsidise costs of travel to conferences because most members have insufficient income to cover such expenses, are faced by disabled women in general. Others, such as hostility and resistance to DAWN's openly feminist politics and to its efforts to promote inclusion of minorities in political action, in particular lesbian women, are specific to DAWN and reflect members' commitment to creating separate spaces for political voice and action by disabled women in Canada (see Stone 1989; Doucette 1991; Driedger 1993; Odette 1993).

With a few important exceptions, most accounts of disabled women's political activism through groups other than DAWN describe experiences within a particular organisation in either the disability or women's movement, or with respect to a particular event concerned with disabled women's issues. Andreychuk (1993), for example, discusses how ten disabled women at the Independent Living Resource Centre in Winnipeg established a support group to deal with issues of abuse in their lives and with barriers they faced to participation in the women's movement. Similarly, Blackford (1993) describes what she terms the 'feminization' of the Multiple Sclerosis Society of Canada, with respect to the issue of motherhood. She points out that women in society chapters, influenced by the women's and disability rights movements, challenged the dominant medicalised model of MS by getting issues such as motherhood on the organisation's agenda. Such accounts provide useful snapshots of disabled women's activism in particular organisations and places.

What these accounts cannot convey is the overall level, geographic scope, foci, and organisational base of disabled women's activism in recent decades. A related limitation of the existing literature is its frequent silence on disabled activism in general and disabled women's activism in particular. For example, Burstow and Weitz's (1988) edited collection on struggles against psychiatric practices in Canada provides little information on disabled women's activism in this context. Indeed, as Driedger (1993: 184–5) notes in her discussion of disabled women's struggles in Canada and internationally:

> In the area of historical research, though, there is very little written by women with disabilities or by others. In the face of the dearth of published information, one needs to turn to primary sources in archives and rehabilitation centres, to look for evidence about disabled women's lives . . . the history of women with disabilities is a vast frontier waiting to be discovered.

My current research based on a survey of disabled women activists across Canada is aimed at addressing some of these gaps in our knowledge about disabled women's struggles.

The few available accounts of the history of disabled women's activism in Canada outside DAWN locate its roots in the emergence of the disability movement in the early 1970s (following quickly on the heels of organising in the United States). From 1972 to 1974 provincial groups representing people with 'cross' or a range of physical disabilities were formed in Alberta, Saskatchewan and Manitoba, in order to provide a stronger political voice and presence for the physically disabled than was possible within organisations based on a single type of disability. By 1976 activists in the western provinces had established the Coalition of Provincial Organizations of the Handicapped (COPOH) which, unlike previous national rehabilitation, professional or charitable organisations, was composed exclusively of disabled people speaking for their rights. Although COPOH was influential in advancing general rights of the disabled in Canada (e.g., through the Canadian Human Rights Act and Charter of Rights and Freedoms), it was clear by the early 1980s that the leadership did not intend to represent disabled women's concerns. Driedger (1993: 175) summarises disabled women's position within the disability organisations of the 1970s in the following way:

> Women with disabilities were included in the memberships and held some positions in organizations of disabled persons in the 1970s. However, their involvement tended to be in supporting roles. Women were committee members and sometimes members of the governing councils of disabled people's provincial organizations. Yet in these roles they often were the carriers of coffee or the workhorses that got committee work done.

Driedger notes that although women were concerned about token representation in some initiatives, disabled women activists believed that they and their concerns were taken seriously by disabled men involved in organisations such as COPOH. However, they learned that this depended on whether the issue in question was of concern to disabled men and on how well disabled women activists acquiesced to the authority of the 'old boys' network'. As Dreidger (1993: 176) puts it:

> But that [being taken seriously] depended on the issues. As long as a woman acted like 'one of the boys' she was accepted. Women's issues were not regarded as 'serious' issues by disabled men in the 1970s and 1980s. As long as women brought up 'important' issues such as transportation, accessibility or housing they were listened to. When disabled feminists Peters and Pat Israel sat on the national council of COPOH and brought up women's issues they were brushed off.

275

Women's struggles to increase their voice and influence in disability organisations during the 1980s were hampered by the political dominance of men. A COPOH national conference resolution supporting the study of women's issues was passed in 1983, for example, but it was 1987 before the organisation acted on the resolution by publishing a discussion paper. This action coincided with the election of COPOH's first female chairperson in 1986. And it was under her leadership (up until 1990) that steps were finally taken to ensure that one-half of the members of the executive were women (Driedger 1993: 180).

The United Nations' declaration of 1981 as International Year of the Disabled provided some interesting opportunities to increase public and government awareness of issues facing disabled women in Canada. Significantly, however, people were not always receptive to the concerns disabled women expressed. The Nova Scotia government, for example, commissioned a disabled woman researcher to document special issues facing women with disabilities. When the Nova Scotia government rejected her final report, based on interview data, as 'too depressing' and having 'too much sex', the Women's Press approached Matthews to write a book based on the report. This led to the 1983 book *Voices From the Shadows*, one of the first Canadian publications to deal with disabled women's issues.

By the mid-1980s, disabled women were making their presence and concerns known within disability and women's organisations. Several disability organisations, including Saskatchewan's Voice of the Handicapped, Toronto's PUSH Central (Persons United for Self-Help) and British Columbia's Coalition of the Disabled had had women's caucuses established. Events such as the Women's Music Festival in Winnipeg in 1984 were providing wheelchair access and offering sign language interpretation of some stage presentations. Sessions on disabled women's issues were being included in conferences, such as the Focus on Women conference in Winnipeg (July 1984). These sessions brought together women active in disability organisations such as the Manitoba League of the Handicapped and community activists concerned with disability issues (Toews 1985). Disabled lesbian women were also organising to get issues of difference and disability on the women's movement agenda. Doucette (1985) reports on a workshop she facilitated on Lesbian Sexuality and the Differently Abled at the Toronto Lesbian Sexuality Conference. She was also asked to join a women's panel discussion on Minorities within Minorities at the annual PUSH conference in Thunderbay. Reflecting on both initiatives, she points out the need to create opportunities for minorities such as lesbians to speak out within the women's and disability movements, and the importance of promoting acceptance of minorities within groups. Doucette notes, for instance, that at the PUSH conference she had greater acceptance as a disabled person than she experienced within the lesbian community. Still other disabled women activists were promoting discussion of disability issues in organisations such as the Canadian Abortion Rights Action League (Israel 1985).

By the 1990s it was clear that disabled women's activism within DAWN and other organisations within the disability rights and women's movements was making a difference. Organisations within both movements were slowly beginning to recognise and address disabled women's concerns. The theme of the 1991 Canadian Research Institute for the Advancement of Women (CRIAW) was women and disabilities. Members of CRIAW and DAWN worked together to ensure that the conference was fully accessible to disabled women. An important feature of the conference was the pairing of each disabled participant with a non-disabled 'sister' to ensure that personal assistance needs were met and to raise awareness of the realities of disabled women's lives (Klein 1996). The conference encouraged feminist academics to begin to tackle disabled women's issues (Israel and Odette 1993). Sessions on disabled women's issues were also becoming more common in disabled persons conferences in Canada. Government officials at different levels were increasingly recognising the importance of supporting initiatives on issues such as violence against disabled women (e.g., through funding provided through the Office for Disability Issues of the Ontario Ministry of Citizenship). Disabled women were being included in consultations with women's groups, for example, through representation on the Ontario Women's Directorate (Israel and Odette 1993). New disabled women's groups were also emerging, a notable example being Voices of Positive Women, an Ontario group representing women disabled by HIV/AIDS (Taylor 1992).

Still, such progress was tempered by funding cuts to organisations such as DAWN, lack of full access to many conferences and feminist academic literatures (e.g., language and print barriers), disabled women's exclusion from many women's services, and exclusionary ableist practices. Israel and Odette (1993: 8) provide the following examples of ableism within the women's movement in Canada:

> Able-ism is also reflected in the kind of language that non-disabled feminists use when referring to feminists with disabilities. For example, 'you are so courageous' or 'it's so nice that you were able to get out and come to this conference.' Able-ism also rears its ugly head when we see that non-disabled women rarely attend the workshops held on disabled women's issues. Recently, at a conference of 300 women addressing the issue of violence, approximately seven women attended the workshop held on disabled women's issues; two in the morning and five in the afternoon. To the numerous women with disabilities who spend many hours preparing these seminars, the low numbers of non-disabled women participating in our workshops may be seen as sending a message about the importance of this issue within the larger context of women's issues.

As the preceding account indicates, disabled women's activism in Canada has faced numerous socio-spatial barriers: from barriers of physical access, distance and poverty, to exclusionary institutions and movements, and ableist attitudes and practices that pervade and constrain the lives of women with disabling differences. In the next section, I look in more detail at the kinds of socio-spatial barriers to political voice and activism that disabled women have encountered in the Canadian context.

Socio-spatial barriers to disabled women's activism

In the best of circumstances, physical distance can be a daunting barrier to political action. However, when one considers the physical separation of disabled activists, especially women, in a country as vast as Canada, the distance barrier can often seem insurmountable. Because most disabled women are too poor to cover travel and accommodation costs from their personal incomes, organisers of meetings, conferences and other events must raise sufficient funds to cover all or most of these expenses. This has been difficult at the regional and provincial levels, and especially problematic at the national level, as DAWN Canada's experiences with funding cuts, for instance, indicates (Dreidger 1993; Briscoe 1993). Overcoming geographic isolation through the use of computer technology is at best a partial solution, since disabled women often lack access to computerised services such as electronic mail. To the extent that such strategies dominate networking and decision-making within organisations such as DAWN or COPOH, they represent technological barriers, and barriers of privilege, to the inclusion of disabled women in political leadership and action.

The spatial and political isolation of disabled women in Canada has been exacerbated by barriers such as physically inaccessible meeting places and lack of accessible transportation. In a ground-breaking 1986 open letter to the women's movement, a letter reprinted in a number of women's publications, DAWN Toronto spelled out what access meant for women with different disabilities and challenged women activists to address these needs. They argued:

> Who would think of putting out a flyer saying: IMPORTANT FEMI-
> NIST EVENT FEATURING MS. DARING DAIRY, WELL KNOWN
> AUTHOR, Nov. 30, 8:00 PM, Everywoman's Hall. Admission FREE.
> Childcare. DISABLED WOMEN NEED NOT APPLY. Of course not!
> Yet often, even usually, that's what the publicity for feminist events
> says to disabled women . . . Your problem is usually that you just plain
> don't know what accessibility is. Our problem is that we can't get in
> to even tell you.
>
> (DAWN Toronto 1986: 80)

The open letter goes on to list different access needs; for example, ramps, adapted washrooms and non-segregated seating for wheelchair users, sign language inter-preters for deaf and hearing-impaired women, printed matter in braille, large print or on audio-cassette and acceptance of seeing eye dogs for visually impaired women, attitudes and language which are sensitive to the needs of developmentally disabled women (e.g., not referring to people as idiots or imbeciles), making sure that the needs of women with invisible disabilities are considered (e.g., providing nutrition breaks for women with diabetes and smoke-free environments for women with environmental illnesses), and provisions for personal attendants when required (e.g., free admission to the meeting or event in question). As the authors note, however, even provisions like these may not make conference or other events accessible to disabled women because of other barriers such as inadequate trans-portation and failure to publicise events within the disabled community:

> Even when events are accessible, you may not see disabled women out. This is often because of transportation. Wheeltrans, Toronto's alterna-tive transit system, is separate from regular TTC, but it is certainly not equal. Wheeltrans users, for example, are not allowed to use Metropasses, but, even though we are the poorest of all women (even old age pensioners get more), we must pay full fare every time. And we must book at least 7 days in advance if we want to go somewhere. And that's no guarantee you'll get there . . . Organizers in the women's movement rarely seem to think of publicizing events in the newspapers or on the phone lines of the disabled movement. If you want us at your event, or in your group, advertise where we read.
>
> (ibid.)

Lack of information about women's events and the absence of non-disabled women at events such as conference sessions on disabled women's concerns, are symptomatic of social barriers to the inclusion of disabled women in the women's movement. Although less 'visible' than physical barriers such as the absence of automatic doors, these social practices of ableism sustain and exacerbate the socio-spatial isolation of disabled women – both within the women's movement and in society more generally. At the personal level, repeated encounters with environments which are both physically inaccessible and socially oppressive have had devastating effects on disabled women activists. In an often-cited article, written in 1985, Pat Israel shared her experiences of being a feminist activist in a women's movement that marginalised disabled women:

> Sometimes I feel like crying and screaming and I just want to quit the feminist movement. But I can't . . . I will always fight for the rights of

women. I just wish that women in the women's movement would recognize us as sisters and instead of putting barriers in front of us would open the doors and welcome us. Years ago when I attended a national women's conference I had to use a dirty, foul-smelling elevator to get to the workshops. There was garbage on the floors and walls. I felt degraded and dirty every time I had to use it. I wonder what would have happened if the black women attending the same conference had been told to use the freight elevator because they were black.

(Israel 1985)

Unfortunately, the personal pain of exclusion and marginalisation expressed by Israel continues to be experienced by many disabled women today. In a 1990 article, Stuart and Ellerington noted that while the disability movement was beginning to pay 'real attention' to disabled women's issues, the women's movement had made much less progress in tackling the exclusion of disabled women. They pointed out that negative attitudes towards disabled women, expressed for example in assumptions that they were uninterested in and incapable of contributing to action on women's issues, helped to ensure that women's groups, events and services remained inaccessible:

Women's organizations seldom consult with disabled women about any issue, they most often hold meetings in inaccessible places, and they fail to provide information in a way that is accessible to all women with disabilities. They often do not recognize that disabled women are interested in women's issues and they assume that our disabilities make us incapable of contributing. Women's services (for example, shelters for battered women) are rarely accessible, creating additional barriers to obtaining the supports and services that other women take for granted.

(Stuart and Ellerington 1990: 16–17)

One of the authors, a survivor of incest and the psychiatric system, explained how lack of understanding of her special needs made the women's movement an unnecessarily threatening environment. Vulnerable to brain seizures and periods of emotional upheaval, she risked rejection within a women's movement that failed to allow for different needs:

If a woman's needs require substantial support in order for her to fully participate, the movement may actively reject her . . . They need to accept my limitations [e.g., withdrawal in social situations] as human rather than bizarre, and provide a safe, nurturing place where I can be vulnerable or strong. They must recognize, as I do, that therapy will probably be an on-going part of my life. They need to know that I could lose everything at any time, including my sense of identity, and need to put the pieces back together once again.

(ibid.: 17)

Experiences such as these remind us that the representation of disabled women within the women's movement, for instance through the National Action Committee on the Status of Women's (NACSW) recently established disability committee, are important but insufficient steps to ensure that disabled women are included within the women's movement. Effective voice and involvement requires recognition that barriers to the disabled affect all our lives, including our understanding of what it is to be 'women'. It requires a willingness to accept disabled women's bodies and minds not as 'less than' but simply as a significant part of the spectrum of differences between people. In short, it requires a feminist body politics which measures its progress in terms of its capacity to value, include, empower and act in solidarity with disabled women.

As indicated above, disabled women in Canada have also had to contend with gendered barriers to effective political voice and action. This has included exclusion from leadership roles within disability organisations, and resistance to hearing about or acting on issues of particular concern to disabled women: such as violence and abuse, and reproductive rights. In this case, disabled female bodies and minds have been treated as 'less than' and 'other' than their male counterparts – capable of supporting political action by disabled men but not of identifying significant political issues and providing the leadership needed for disabled activists to tackle them. The extent to which disabled women have effectively challenged patriarchal oppression within the disability movement remains unclear. On the one hand, there are scattered reports in the existing literature of particular activists saying that significant progress has been made in getting women's issues on the disability community's political agenda. On the other hand, there has been sufficient dissatisfaction with disabled women's marginalisation within the disability community to give rise to a separate disabled women's movement in the form of DAWN and through disability caucuses or committees within organisations such as NACSW. The outcome of efforts to establish a women's caucus within COPOH provides an instructive example of the difficulties disabled women have encountered in trying to create spaces of resistance and action within the disability movement. Male COPOH members insisted on and won the right to membership in the caucus, thus making it impossible for disabled women to deliberate on issues of concern without male presence or intervention within the formal structure of the organisation. This in turn meant that disabled women would have to try to develop their political voice, agenda and influence in contexts still permeated by male privilege and power.

281

Disabled women and the state

It is important to recall that Canada's disability and women's movements have developed within an ableist, sexist, heterosexist and capitalist society. And as significant as attitudes and practices that exclude disabled women within each movement are, these remain sites of resistance to at least some facets of disabled women's oppression. In this section, I examine socio-spatial barriers to disabled women's well-being and political activism associated with state restructuring and social policy changes. I argue that disabled women are caught up in a regulatory process that seeks to limit and discipline the use of state resources by disadvantaged groups, and which perpetuates the socio-spatial isolation of persons with disabling differences from the political process. Disabled women activists are, in this context, experiencing growing socio-spatial barriers to daily life and political action within a political system of 'apartheid' based on the devaluation of persons and especially women with disabling differences.

Recent changes in state policy towards the disabled at the federal level of the Canadian state and at the provincial level in Ontario suggest the emergence of a regulatory regime characterised by the dismantling of programmes to assist the disabled, exclusion of the disabled from the policy process, resistance to pro-active legislation to protect the rights of the disabled, reductions in social assistance, and disciplinary measures and incentives aimed at reducing the numbers of disabled persons receiving welfare benefits.

In 1996, the federal Minister of Human Resources indicated Liberal government plans to wind down support for disabled Canadians. Steps towards this end have included major reductions in government support for Medicare and termination of the National Strategy for the Integration of Persons with Disabilities. Recent changes in the federal Income Tax Act have decreased the number of disabled Canadians who can qualify for the Disability Tax Credit (by an estimated 10,000) by making the credit conditional on income from employment. This leaves an estimated 2 million disabled Canadians ineligible for tax credits that could assist with expenses such as mobility aids and renovation of living environments (McDonough 1997). These developments promise to worsen the economic situation of disabled Canadians, and create growing financial constraints on their ability to access community services and participate in political action. In addition, as indicated above, reductions in federal support for disability organisations such as DAWN have made it extremely difficult for disabled women to maintain national offices and networking, and to organise events such as national meetings and conferences. Termination of a national strategy for the integration of the disabled not only eliminates opportunities for federal leadership in areas such as promoting employment of the disabled, but also signals that the Canadian government is not prepared to pressure provincial governments to address disability issues in a consistent way (e.g., through national standards on matters such as access to education and social assistance or accommodation of the disabled in the workplace), or to build upon the legal protections from discrimination provided in the Canadian Charter of Rights and Freedoms (through, for instance, enactment of a Canadians with Disabilities Act). Federal retreat from disability issues and rights also leaves disabled Canadians extremely vulnerable to provincial government actions, such as reducing funding for community care programmes, at a time when many social programmes are already reeling from provincial funding cuts.

In Ontario, the Conservative government record on disability issues is as dismal as it is discouraging to the disability movement and its supporters. In 1995, premier Harris made an election promise that his government would enact an Ontarians with Disabilities Act within his first term of government, would work with a broad-based coalition of legal experts and community activists (known as the Ontarians with Disabilities Act committee) to develop the legis-lation, and would provide new funds to promote accommodation of the special needs of persons with disabilities. Since that time, the premier has repeatedly refused to meet with the coalition, the responsible Minister met with them only once (in 1996), and no official action on either matter has been taken. This is despite the passing of a 1996 resolution in provincial parliament supporting a disability act. The 1996–97 business plan of the Ministry of Citizenship, Culture and Recreation (responsible for disability issues) made no reference at all to the Act. As the 'achievements' section of the 1997–98 business plan makes clear, the ministry's priorities have been spending and staff reductions: in 1996–97, spending was reduced by $64 million (17 per cent) and ministry staff by 32 per cent (Ontario Ministry of Citizenship, Culture and Recreation 1997). The ministry's only action on disability issues was an expenditure of $1.5 million to promote employment of persons with disabilities (part of a general programme to promote equal opportunities for employment).

When over 200 disability activists held a 'face the disabled community' session at the provincial parliament to discuss these and other issues with government representatives in May 1997, no Conservative members of parliament were present. This prompted strong criticism from opposition members during the session of parliament that followed. A Hamilton East MPP, Dominic Agostino (member of provincial parliament), for example, noted that representatives of both opposition parties (Liberal and New Democratic) attended and spoke to participants. He went on to observe that:

> What was disgraceful about today was that not one of the eighty-one members of the government side had the guts to face the [disability] community, not one of you.
>
> (Agostino 1997)

To make matters worse, physical access problems prevented the activists from observing parliamentary proceedings in person (e.g., the public gallery only accommodated four wheelchair users), and they were forced to observe debates through television monitors scattered throughout the parliament buildings. For those involved, this exclusion from public space was a frustrating and telling indication of barriers to full rights of citizenship. This dispersal and invisibility also decreased the impact of the protest action within parliament and the press. For example, when an opposition member reminded the Minister responsible for disability matters that a large group of disabled citizens had gathered to protest about the government's inaction, the Minister responded with 'what crowd?' and gestured towards the near-empty visitors' gallery (Ontario Hansard 15 May 1997).

MPPs report other socio-spatial barriers to the participation of the disabled in policy debates and deliberations within provincial government buildings: such as locked doors preventing disabled committee members from accessing rooms used for evening meetings and lack of staff to assist persons with disabilities in finding accessible routes or accessing food and other services (ibid.).

The current provincial government, dedicated to continued reductions in social spending (with the short-term objective of delivering promised tax cuts), has implemented a wide range of funding, policy and programme changes which threaten the well-being of the disabled in Ontario. Cuts to paratransit services have been implemented across the province, for example, leading to situations where persons with disabilities are unable to access transit for pressing medical services and are isolated for longer and longer periods within their homes (e.g., recent reports in Hamilton indicate a waiting period of eight weeks or more for paratransit services) (McMaster University Summer Gerontological Institute Proceedings June 1997). Funding for home care services has also been cut, reducing the number of disabled persons who receive home care and, for those who continue to do so, resulting in major reductions in hours of assistance. In a recent case discussed in provincial parliament, a daughter and son-in law attempting to care for an 81-year-old woman suffering from multiple sclerosis and bedridden in their home were told that home care services would be terminated. The couple planned to appeal against this decision but the only tribunal in place to hear their case (a tribunal established by the current government) was not technically charged with hearing such appeals and government lawyers were preparing to argue that the appeal should not be heard (Ontario Hansard 15 May 1997). The outcome of the appeal was not known at the time of writing, although government representatives claimed that home care would be provided until a decision was reached.

Other provincial actions which threaten to heighten the isolation and exclusion of the disabled in Ontario include: attempts to decentralise responsibility for social programmes such as long-term care to municipalities (which lack the funds to support them), taskforce recommendations that provincial human rights legislation be altered so that the duty to accommodate the disabled (e.g., in the workplace) is substantially weakened, reductions in the funding of the Human Rights Commission which are reported to have resulted in an unprecedented number of disability discrimination cases being turned away, and abolition of a fund within the Ontario public service that was devoted to removing systemic barriers facing disadvantaged groups such as the disabled (Ontario Hansard 15 May 1997). As noted above, the province has also reduced funding to organisations such as DAWN Ontario; in DAWN's case forcing the organisation to move from Toronto to North Bay in order to reduce costs. While this location has the potential advantage of improving access to DAWN in northern Ontario, it also means that these disabled women activists no longer have immediate access to the many disability organisations headquartered in Toronto or to the provincial government.

The state, then, at the federal level and in Ontario, is a rapidly shifting and in many ways increasingly exclusionary terrain of struggle for disability activists. This is especially true for disabled women who must overcome particularly severe barriers to participation in the political process: including extreme poverty, political marginalisation of women within the disability movement, and ableism within the women's movement. And having the fewest resources to begin with, it is also disabled women who are likely to suffer most from the development of a harsher, more disciplinary regulatory regime within the state. Less likely to be employed than disabled men, women with disabilities are most likely to be affected by federal restriction of disability tax credits to earned income, for example. This in turn means that disabled women may have to 'do without' needed assistive devices, which will further increase their isolation and exclusion. This is made especially likely as a result of the termination of provincial programmes to subsidise such purchases (e.g., as is the case in Ontario). Similarly, provincial programmes such as 'workfare' (intended to remove people from the welfare rolls) are likely to pose particular hardships for disabled women judged capable of work; for instance, with respect to wage levels and the double burden of ableism and sexism in the workplace. Of course, current processes of state restructuring promise to have harsh consequences for disabled men as well – the point here is that in the general context of extreme marginalisation of persons with disabling differences, it is disabled women who tend to fare worst of all.

Getting global: international organising initiatives

If there is a 'silver lining' to be found within the circumstances of disabled women's lives and political struggles, it is that disabled women continue to fight back and to mobilise for social change. This is true not only in advanced capitalist countries such as Canada and Britain, but also in developing and Third World nations. Disabled women have also begun to mobilise at the international

scale; although such efforts continue to be hampered by lack of funds and difficulties in organising an extremely marginalised and devalued constituency. In
this section, I discuss some of these global struggles 'against all odds'.

Since the mid-1980s, disabled women have been struggling for a political voice
both within existing international organisations of the disabled and through efforts
to establish separate international organisations of disabled women. By 1985, disabled women were actively challenging the leadership within the World Coalition
of Disabled Women (a global umbrella organisation for disability activism) to
launch reforms that would make the organisation more inclusive of disabled
women and disabled women's issues. In July of the same year, a workshop on
barriers to women with disabilities was held as part of the World Conference to
Review the Achievements of the United Nations Decade for Women in Nairobi,
Kenya. Through this workshop, women from fourteen countries called on the UN
to ensure that disabled women had full access to future conferences and on national
governments to ensure that disabled women were able to participate in all facets
of life. A loose international network of disabled women activists, known as
Disabled Women International, was also established through the workshop.
Unfortunately, lack of funding has prevented the network from doing much more
than producing an annual newsletter (Driedger 1993).

In 1990 a second global disabled women's organisation was established: Disabled
Women International. This organisation was launched at the Vienna United Nations
meeting of Disabled Women. The initiative was prompted by disabled women's
concerns about continued exclusion within both the disability and women's movements, and the relative inaction of Disabled Women International (ibid.).

Efforts to promote disabled women's rights and well-being at the global scale
have been hampered not only by lack of funding, but also by the geographically
uneven development of disabled women's activism. Separate disabled women's
organisations have been concentrated in relatively affluent countries (notably
Australia, the United States and Canada), with the exception of the Uganda
Disabled Women's Association which combines self-help through the sale of crafts
with recruitment of women in rural areas (ibid.). Such separate organisations have,
at least arguably, several potential advantages over activism within the established
disability and women's movements as a 'seedbed' for global organising. These
include: creating international spaces of action in which disabled women can concentrate on issues most pressing to them, nurturing an international political presence for disabled women, providing an organisational base for fund-raising
devoted exclusively to disabled women and for channelling those funds to groups
that prioritise disabled women's groups and projects, nurturing disabled women's
own expertise and leadership, and developing a global political agenda which
reflects the multiple needs and oppressions of disabled women world-wide. The
absence of such separate organisations in many countries, in turn, makes it difficult

to develop the 'critical mass' (e.g., of leadership or funds) needed to support global organising efforts and networks. At the global scale, barriers such as distance, differences in cultural and economic oppression, and poverty, take on much more gargantuan proportions as obstacles to effective political action, making it crucial to have national and regional political bases on which to build.

Disabled women are also continuing to mobilise through established global organisations and networks. Efforts to make Disabled Peoples' International more inclusive of disabled women at its second World Congress, in the Bahamas in 1985, led to the establishment of the DPI Women's Committee. This committee is devoted to working for the special interests of disabled women and girls through the DPI. As part of this work, it also monitors disabled women's involvement within the organisation and outside with the aim of ensuring that development initiatives launched through various organisations prioritise disabled women's needs. The committee is actively expanding its linkages with other international disability and women's organisations including: the World Blind Union, the World Federation of the Deaf and the Women's Environment & Development Organization.

Like other disabled women's groups, the DPI Women's Committee's mandate (c.1985) is an ambitious one:

> To increase the participation of women in their communities and within disabled people's organizations, including DPI. To raise international awareness of the issues affecting women with disabilities, including human rights violations. To maintain a worldwide communications network via newsletters. To gather research material on the specific issues faced by women with disabilities in various regions and countries. To promote and monitor the involvement of women in DPI's organizational structure and in DPI activities. To prepare documentation and presentations for the World Congress/World Assembly.
> (DPI Women's Committee 1998)

The Committee cites many achievements. Within DPI, women's involvement in leadership roles has increased at global, national and regional scales (e.g., several national chapters of DPI now have women's committees and the current executive director of DPI is a woman). The committee has an official presence and voice at DPI World Congresses and workshops and other events sponsored by DPI are paying greater attention to disabled women's issues. Information-sharing amongst disabled women from various countries is facilitated through activities such as the committee newsletter and seminars. Networking with other organisations, such as the UN, UNIFEM, Global Fund for Women, and Oxfam has encouraged greater interest in and attention to disabled women's issues (ibid.).

As the initiatives discussed here indicate, disabled women are taking their struggles into the global arena. In doing so they are not only challenging ableism in its multiple and diverse forms world-wide, but are also helping to forge alliances within the global community of disabled women that begin to transcend boundaries of nation-states, diverse cultures and differences in economic well-being and opportunities. Unfortunately, however, we still know far too little about such struggles and alliances, and about who they empower and who remains excluded. Clearly, in one way or another, disabled women activists who are able to participate in organising efforts at events such as the UN Women's Conference in Beijing (e.g., members of the DPI women's committee) are a relatively privileged minority: with connections to others in power within organisations such as the UN, in positions of leadership within disabled women's organising within their own countries, and able to tap the economic and other resources needed to travel and interact with others at the global scale. It is also unclear how well such women represent the range of disabilities that women cope with world-wide. It would be interesting, for example, to investigate whether and to what extent women with disabilities that result in difficulties communicating in public are represented as participants in events such as panel discussions or policy debates or in leadership positions within international groups and networks. To what extent are other differences, in sexual orientation for instance, barriers to inclusion of disabled women in international organising initiatives?

Answers to questions like these will require concerted efforts to tell the stories of disabled women's activism at the global scale in its full richness and diversity. And it will require that activists and researchers ask some 'hard' questions of international struggles to advance disabled women's rights and well-being; including questions about inclusion, exclusion and silencing. With so many odds stacked against disabled women, it is sometimes tempting to shy away from such difficult issues and to romanticise any and all organising efforts. As the Canadian experience suggests, however, disabled women and their allies cannot afford such luxuries: the barriers to inclusion and social change are as much within as without disabled women's struggles.

Conclusions: making diverse bodies matter

The developments discussed in this chapter make three things quite clear. First, that in recent decades disabled women have struggled for and won increased influence within the disability and women's movements, and have also engaged in separate struggles through independent disabled women's organisations and networks. These are especially hard-won victories from the 'margins of the margins'; the results of struggles waged under extremely disabling conditions of poverty and exclusion. Moreover, as some of the personal experiences cited in this chapter indicate, disabled women's political gains have often come at high personal costs: from struggling to reconcile a commitment to feminism with the realities of exclusion from women's events, to internalising messages that a disabled female body and mind are somehow 'less worthy' than others. Disabled women's struggles for political voice and action have also made a real difference in women's lives. Bonnie Klein (1996), in a moving account of her struggles to recover from a debilitating brain stem stroke, likens her discovery of the disabled women's movement (in Canada and internationally) to her earlier introduction to feminism in terms of finding spaces in which she was no longer 'other' and reawakening to possibilities for promoting social change through political action. Even in the face of an exclusionary political process, the fact that disabled women are organising for social change can be a lifeline to women trying to come to terms with the social consequences of disabling differences.

The second conclusion to be drawn from this chapter is that recent state restructuring is, in Canada and in Ontario, tending to reduce state involvement in disability issues, restrict access to disability-related programmes and, through expenditure reductions and down-sizing, increase multiple barriers to the participation of the disabled in the political process (e.g., increasing geographic isolation through cuts in transit services and reducing the capacity of government bodies such as the Human Rights Commission to deal with cases of discrimination on the basis of disability). This paring down and shedding of state support for the disabled includes direct reductions in government funding for organisations promoting the rights of persons with disabling differences – heightening constraints on organising activities.

The regulatory regime emerging within the Canadian state is perhaps best characterised as one of enforcing increased socio-spatial marginalisation of the disabled (an apartheid ableism), disciplining the demands of the disabled upon state resources, and increasing employment incentives (for all those receiving state income support, including the disabled). This is an extremely ableist regulatory regime: providing special benefits for disabled persons able to secure and hold jobs, and prioritising the well-being of able-bodied and relatively affluent citizens (e.g., through major tax cuts) over the needs of disadvantaged groups. In this context, token government actions, such as the Harris government promise to consider a provincial act to protect the rights of the disabled, ring very hollow.

Finally, it is clear that disabled women have 'arrived' on the global political scene and are slowly establishing the alliances needed to nurture an international basis for political action. Their efforts are hampered not only by lack of funding but also by multiple barriers of diversity with which such organising must inevitably contend. The absence of separate disabled women's organisations in most countries presents further challenges, including limiting opportunities for fund-raising and for channelling funds to political projects aimed at empowering disabled women.

Still, much has and is being accomplished under the auspices of organisations such as Disabled Peoples' International (DPI). As is the case with political action at national and local scales, it is important that disabled women active in the international political arena be alert to bases of exclusion within their own organising efforts and work to build an inclusionary and representative global movement.

What sorts of 'body politics' can disabled women activists use to unsettle oppressive regimes of power associated with phenomena such as state restructuring and with barriers related to differences amongst disabled women? One possibility, now that the different concerns of disabled women are better recognised and accepted within the disability and women's movements, is a campaign to network with an array of disadvantaged groups negatively affected by social policy change (e.g., anti-poverty groups, immigrant groups). This could help encourage an emphasis on a politics of solidarity in diversity; and complement disabled women's established emphasis on developing distinct political voices and platforms. It could also encourage non-disabled and disabled male activists to recognise common stakes in state restructuring – despite differences in body status and gender. To recognise, for example, that lack of provision for the needs of the disabled harms all citizens, be it through the loss of disabled activists' insights regarding the human costs of state restructuring or more personally with respect to inadequate care for a loved one. Of course, such coalitional initiatives require resources which have already been severely cut back. A potential strategy to help overcome resource constraints is to exploit more fully the potential of computerised communication – in particular, in ways that reach disabled women and other groups currently without access to 'virtual politics'. This might include initiatives to provide computer access through community centres and local disability organisations for the explicit purpose of tracking policy change and collaborating on political strategies.

Outreach to disabled women currently without access to other women activists and activist organisations is another way in which disabled women's capacities to challenge oppressive state regulatory regimes can be increased. A possible strategy in this regard is a 'big sister' programme where established activists sponsor particular individuals and groups by sharing information on policy developments, helping to identify local organisations concerned with policy issues, and perhaps even sharing expertise in establishing local organisations and linking those organisations to others at local, national and international scales. In the Canadian context, this could include establishing an interactive national DAWN website that provides regular updates on political and legal issues affecting disabled women, provides a forum for discussion, and offers direct assistance to disabled women involved in (or contemplating) political organising.

Another key challenge for disabled women activists and others in Canada is to publicise the human costs associated with the regulatory regime emerging within the Canadian state. Informational pickets, for instance at the provincial parliament and community facilities used by disabled citizens, could be used to publicise lack of access to the political process and the impacts of cuts to services such as paratransit to disabled persons' access to community services and spaces. Other complementary strategies include computerised information-sharing on the impacts of policy change and local efforts to network with members of the mass media in order to heighten awareness of the serious consequences of cutbacks to programmes such as home care. Organisations such as the National Action Committee on the Status of Women, which has recently established a disability committee, could also be enlisted in publicity work.

These kinds of initiatives are also needed to strengthen and build upon disabled women's activism internationally. Disabled Women International could, with sufficient resources, become a resource centre for global action: tracking developments in state policies affecting the disabled, sharing key advances in disabled women's rights and well-being, publicising opportunities to participate in disabled women's struggles, and supporting efforts to increase access to the disabled women's movement through outreach and advocacy. Current initiatives, such as DPI's women's committee's efforts to recruit disabled women representatives from countries not currently represented within DPI, could provide the basis for an expanded international 'big sister' programme in which established women's organisations sponsor the development of disabled women's groups in under-represented regions and countries. The latter could include projects directed at increasing disabled women's access to information and strategic debates through computer technology; working through development agencies, for example, in order to provide local community centres and disabled women's groups with access to internet resources. A related challenge is to make more disabled and non-disabled women aware of the issues and challenges facing disabled women globally and locally. Established groups within the women's movement, although still often exclusionary in practice, are becoming aware of and increasingly allied with disabled women's struggles and can help to further those struggles through advocacy, fund-raising and information-sharing. With so few resources to draw upon, the global disabled women's movement survival, influence and impact will depend upon building strong and inclusive alliances with other women's groups and disability activists.

So far I have concentrated on some of the practical ways that disabled women activists can strengthen their 'body politics' in the Canadian and global contexts. Are there also shifts in philosophical or conceptual perspectives that might strengthen disabled women's capacities to unsettle oppressive regimes of power? To date, disabled women activists have, albeit to varying degrees, embraced a politics of body difference. This has sometimes taken explicit feminist form, for example, in DAWN's mandate to empower women with disabling differences or the creation of a disability group within NACSW. In other situations, for

291

instance within COPOH or DPI, it has taken the form of insisting that the disability movement take issues of concern to disabled women seriously by putting them on the political agenda. As indicated above, these efforts have helped to provide disabled women with a sense of political identity, safe spaces for political discussion, and growing awareness of the possibilities for social change through political action. One way to build on these achievements might be to conceptualise ableist and sexist socio-spatial barriers as the product not only of discrimination on the basis of bodily differences, but also of a more general cultural and societal intolerance of bodily diversity. This amounts to a kind of corporeal 'class system' in which the value assigned to particular kinds of bodies by powerful groups such as capitalists depends upon whether and how these bodies can be exploited in order to further the accumulation of wealth. In this system, for example, a premium is placed on bodies that can project Western cultural ideals of beauty and success through advertising and other media, and fuel international consumption of commodities such as movies, clothing lines and perfumes. Those who excel at servicing the image needs of corporate capital, either through deploying their own bodies or managing the bodies of others, may even rise to the ranks of highly paid 'stars' within the film and fashion industries for example. Bodies marked by differences such as race, culture, age or disability, on the other hand, occupy the lower rungs of status within the corporeal class system, and are especially vulnerable to economic and social exploitation and marginalisation. Manifestations of these processes of corporeal oppression include: the exploitation of Asian women and children in the international sex trade, the hyper-exploitation of female workers in Third World branch plants producing commodities such as automobile parts and computer chips, and the (ab)use of disabled persons' labour in sheltered workshops (e.g., little pay and poor working conditions). Related regimes of power, such as compulsory heterosexuality, compound these bodily oppressions, in this case by disciplining and penalising women and men deemed to use their bodies and perform their gender roles in transgressive ways. In all these socio-spatial contexts it is, of course, the powerful who decide which bodies matter, from the corporate executives and branch plant managers who decide what Mexican or Asian female labour is 'worth' in terms of pay and working conditions, to fashion and advertising moguls who determine which faces and bodies 'sell' and when profitability demands that even these corporeal images be replaced (e.g., the discarding of flagship faces and models by corporations such as Lancome when they reach the age of forty), and even to government leaders who help to determine which bodies will have access to rights and services and which will not.

Conceptualising the body as a general and diverse site of oppressions in late patriarchal capitalism helps underscore the connections between such apparently different phenomena as the devaluation of disabled women in Canada and the hyperexploitation of female workers in Mexican branch plants. In the former case, female bodies and minds which are impaired are judged 'inferior' and incapable of societal contributions (e.g., through employment and culture). In the latter case, female bodies and minds which differ from those highly valued in countries in which corporations are headquartered are devalued by reducing their significance to their capacity to labour. Both are forms of intolerance to corporeal diversity which treat female bodies merely as a means to the end of the accumulation of wealth, failing to appreciate familial or other dimensions of these women's lives or how conditions such as excessive working hours, poverty and malnutrition limit the societal contributions these bodies can make (not to mention exacerbating personal suffering). When we begin to think about corporeal oppression in these terms it becomes possible to see phenomena such as the hyper-exploitation of female workers' bodies in Third World branch plants as one of the 'flip sides' of the socio-spatial marginalisation of disabled women in first world countries. In the former case, wealth is drained from the host country not only through profits but also through exhausting women's capacities to labour as well as their health. In the latter case, social exclusion and poverty drain disabled women's mental and physical capacities to contribute, thus robbing these women of opportunities and society more generally of their many potential contributions.

Clearly, a feminist geographic theory of the body as a multiple and diverse site of oppression and resistance offers one promising intellectual basis for coalitional political struggles around disability issues at various scales. Whether such a complementary political initiative can be realised as state restructuring and economic marginalisation starves disabled activists of organising resources remains to be seen. What is clear is that disabled women's activism is alive, if not altogether 'well', and that disabled women and their supporters are determined to make our bodies 'matter'.

References

Agostino, D. (1997) Statement on Accessibility for the Disabled, Ontario Hansard, 15 May available at: http://www.ontla.on.ca/hansard/hansard.html.

Andreychuk, T. (1993) 'Sharing our expertise through peer support', *Canadian Woman Studies* 13, 4: 99–101.

Barile, M. (1993) 'Disabled women: an exploited underclass', *Canadian Woman Studies* 12, 4: 32–3.

Blackford, K. (1993) 'Feminizing the Multiple Sclerosis Society of Canada', *Canadian Woman Studies* 13, 4: 124–30.

Briscoe, S. (1993) 'DisAbled Women's Network: cuts threaten dusk for DAWN', *Kinesis* April: 3.

Burstow, B. and Weitz, D. (eds) (1988) *Shrink Resistant: The Struggle Against Psychiatry in Canada*, Vancouver: New Star Books.

DAWN Canada available at http://www.indie.ca/dawn/index.html.

DAWN Ontario (1997) February and July Newsletters (available at: http://www3.sympatico.ca/odell/dawnpage.html).

DAWN Toronto (1986) 'An open letter from the DisAbled Women's Network, DAWN Toronto to the Women's Movement', *Resources for Feminist Research* 15: 80–1.

Doucette, J. (1985) 'Breaking the links of lies', *Resources for Feminist Research* 14, 1: 9–10.

—— (1991) 'The DisAbled Women's Network: a fragile success', in J. Wine and J. L. Ristock (eds) *Women and Social Change: Feminist Activism in Canada*, Toronto: Lorimer, 221–35.

DPI Women's Committee (1998) 'DPI Women's Committee homepage' (available at: http://www.escape.ca.women.html).

Driedger, D. (1993) 'Discovering disabled women's history', in L. E. Carty (ed) *And Still We Rise*, Toronto: Women's Press, 173–87.

Israel, P. (1985) Editorial introduction, *Resources for Feminist Research* 14, 1: 1–2.

Israel, P. and Odette, F. (1993) 'The DisAbled Women's Movement 1983 to 1993', *Canadian Woman Studies* 13, 4: 6–10.

Klein, B. Sherr (1992) 'We are who you are: feminism and disability', *Ms. Magazine* 3: 70–4.

—— (1996) *Slow Dance: A Story of Love, Stroke and Disability*, London: Random House.

Matthews, G. Ferguson (1983) *Voices From the Shadows: Women with Disabilities Speak Out*, Toronto: Women's Educational Press.

McDonough, A. (1997) 'Equality for persons with disabilities', commentary by Canadian NDP leader (available at: http://www3.sympatico.ca/alexa.halifax/disablefact.html).

McMaster University Summer Gerontological Institute Proceedings (1997) Special session on Advocacy and Disability Research, June, Hamilton, Ontario: McMaster University.

Odette, F. (1993) 'Women with disabilities: the third "sex" – the experience of exclusion in the movement toward equality', unpublished independent inquiry project in partial fulfilment of requirements for the Masters of Social Work degree, Faculty of Social Work, Carleton University.

Ontario Hansard, 15 May 1997, members debates on Access for the Disabled (available at: http://www.ontla.on.ca/hansard/hansard.html).

Ontario Ministry of Culture, Citizenship and Recreation (1997) *1997–98 Business Plan*, Toronto: Ontario Ministry of Culture, Citizenship and Recreation.

Statistics Canada (1995) *A Portrait of Persons with Disabilities*, Ottawa: Statistics Canada.

Stone, S. D. (1989) 'Marginal women unite! Organizing the DisAbled Women's Network in Canada', *Journal of Sociology and Social Welfare*, 16, 1: 127–45.

Stone, S. D. and Doucette, J. (1988) 'Organizing the marginalized: the DisAbled Women's Network', in F. Cunningham *et al.* (eds) *The Politics and Practice of Organizing*, Toronto: Between the Lines, 81–97.

Stuart, M. and Ellerington, G. (1990) 'Unequal access: disabled women's exclusion from the mainstream women's movement', *Women and Environments* 12, Spring 1990: 16–19.

Taylor, D. (1992) 'Women, Aids and disability issues', *ARCH-TYPE* Jan./Feb.: 13.

Toews, C. (1985) 'Two battlefronts: the Women's Movement and the Disabled Consumer's Movement', *Resources for Feminist Research* 14, 1: 1–9.

294

INDEX

Note: **emboldened** page references refer to chapters